Th

Manufacturing Planning and Control

Manufacturing Planning and Control

Beyond MRP II

Paul Higgins

University College Galway
Galway
Ireland

Patrick Le Roy

SAP Belgium
Brussels
Belgium

and

Liam Tierney

Digital Equipment
Dublin
Ireland

CHAPMAN & HALL

London · Glasgow · Weinheim · New York · Tokyo · Melbourne · Madras

Published by Chapman & Hall, 2–6 Boundary Row, London SE1 8HN, UK

Chapman & Hall, 2–6 Boundary Row, London SE1 8HN, UK

Blackie Academic & Professional, Wester Cleddens Road, Bishopbriggs, Glasgow G64 2NZ, UK

Chapman & Hall GmbH, Pappelallee 3, 69469 Weinheim, Germany

Chapman & Hall USA, 115 Fifth Avenue, New York, NY 10003, USA

Chapman & Hall Japan, ITP-Japan, Kyowa Building, 3F, 2-2-1 Hirakawacho, Chiyoda-ku, Tokyo 102, Japan

Chapman & Hall Australia, 102 Dodds Street, South Melbourne, Victoria 3205, Australia

Chapman & Hall India, R. Seshadri, 32 Second Main Road, CIT East, Madras 600 035, India

First edition 1996

Typeset in 10/12 pt. Zapf Calligraphic by Saxon Graphics Ltd, Derby
Printed in Great Britain by St Edmundsbury Press, Bury St Edmunds, Suffolk

ISBN 0 412 55300 7

A catalogue record for this book is available from the British Library

Library of Congress Catalog Card Number: 95-83096

Printed on permanent acid-free text paper, manufactured in accordance with ANSI/NISO Z39.48–1992 and ANSI/NISO Z39.48–1984 (Permanence of Paper).

Contents

Foreword

Today's competitive manufacturing environment places tremendous pressures on manufacturing plants. Demanding customers require increasingly customized products of the highest quality and at the lowest price, delivered just in time. Such demands stretch the limits of today's manufacturing systems technology. MRP-style systems, conceived in an era of production-led and stock-based manufacturing, have been augmented by JIT thinking but nevertheless have difficulty in coping with the demands of manufacturing.

The pressures on manufacturing planning and control systems are felt most keenly in the areas of business planning, master scheduling and shop floor control. The 'middle' level, namely requirements planning, works reasonably well. This book is particularly strong in the areas of business planning and master production scheduling. Furthermore, the authors succeed in discussing these topics in the context of the different manufacturing environments, including MTS, ATO and MTO.

The book arises from work carried out within the European Strategic Programme of Research in Information Technology (ESPRIT), and in particular in the project IMPACS (integrated manufacturing planning and control systems). IMPACS brought together leading-edge companies and university laboratories from Europe and gave them the opportunity to work together on the topic of manufacturing planning and control. IMPACS was an industry-led project, as will be evident to readers of this book, and at all times sought to relate state-of-the-art theory and the latest research thinking to best manufacturing practice.

In fact, it is this almost unique combination of 'theory tested in practice' that makes this book so worthwhile, in my view. The authors, who come from diverse backgrounds, including research, industry and manufacturing systems consultancy, have taken advantage of IMPACS to develop their ideas and articulate them clearly in this book.

I have no hesitation in congratulating the authors on their efforts and in recommending their book to manufacturing systems researchers and practitioners. I am convinced that it represents a valuable addition to the literature on manufacturing planning and control.

Professor Jim Browne
Director – CIMRU
Galway
May 1995

Acknowledgements

We would like to begin by thanking all the people from whom we have learned about manufacturing planning and control, through the many different discussions among ourselves and our promoters and with numerous industrial and academic colleagues over the last few years. These discussions have actually created what the book has become. We would like to express our deepest appreciation and thanks to Professor Jim Browne from CIMRU, University College Galway (UCG), Ireland, and Roger Packer from PA Consulting, who have both been great supporters throughout the entire preparation time of this book. We also sincerely thank all our colleagues, in particular Luc Delombaerde, Keith Smurthwaite, Luc Chalmet, Bernard Sauve and Franck Dupas, who were a source of inspiration, and the European Commission experts who reviewed our work. In addition, we gratefully acknowledge Martin Gallagher of the Eastern Health Board, Ireland for his ongoing support. We would like to thank Mark Hammond and all the people at Chapman & Hall who were extremely patient in waiting for this manuscript.

Finally, we have learned how long the writing of a book can take and how much work is required not only by the writers themselves, but also by their family members. Our families were very supportive and their patience was often tested. We dedicate this book to Martynn, Hilde and Niamh.

About the authors

Paul Higgins

After studying industrial engineering (1986) at University College Galway (UCG), Ireland, Paul Higgins then worked on the ESPRIT projects COSIMA and IMPACS while studying for his masters degree (1988) and doctorate (1991) at UCG. This work was in the area of computer-integrated manufacturing with special emphasis on production planning and control. Since receiving his doctorate, he has worked on numerous other consultancy and research projects in the area of production management and information technology. He is currently lecturing in information technology at UCG.

Patrick Le Roy

Patrick Le Roy has a master's degree in engineering (1986) from the State University of Ghent (RUG) as well as a Bachelor's degree in economics (1987) and a degree in business management (1993) from the Free University of Brussels (ULB). He also has a CPIM (1995) from the American Production and Inventory Control Society. He gained experience in the domains of computer-integrated manufacturing and logistics with, among others, the State University of Ghent and as a consultant with PA Consulting Group. Within PA Consulting Group he was one of the principal actors active in the ESPRIT project IMPACS. He is now working as a logistics systems consultant for SAP Belgium.

Liam Tierney

Liam Tierney has a degree in industrial engineering (1983) from University College Galway. He has a CPIM (1984) from the American Production and Inventory Control Society and has lectured for the Irish Production and Inventory Control Society. He has 10 years of practical experience working in discrete manufacturing with Digital Equipment, holding various engineering and materials management positions. Within Digital, he was one of the principal persons active in the ESPRIT project IMPACS. He is currently working as a management consultant with Digital concentrating on planning systems and business process re-engineering.

An introduction to manufacturing planning and control systems 1

1.1 INTRODUCTION

There has been much written about the many different manufacturing planning and control philosophies that are used in manufacturing. The advantages and disadvantages of each approach are outlined and discussed in numerous technical papers and books. One of the main aims of this book is to prove that the specifics of the manufacturing environment should be taken into account when designing and implementing a manufacturing planning and control system.

Many companies have adopted the approach of material requirements planning (MRP) and manufacturing resource planning (MRP II). This has often been done without considering the manufacturing environment in which these companies operate. An ideal planning architecture may involve the use of MRP II, in which the main focus is on the planning of materials, or another approach in which more attention would be on the planning of available capacity. Despite the improvements and broadening of the MRP framework, MRP II systems still appear to perform poorly in certain manufacturing environments. This poor performance makes it necessary to review the underlying assumptions in order to be able to evaluate their applicability and develop planning and control architectures that are tailored to the needs of each individual company.

With the arrival of flexible manufacturing systems (FMS) in the 1970s and computer-integrated manufacturing (CIM) systems in the 1980s, a new type of manufacturing environment is beginning to emerge. This new environment is one in which total customer satisfaction is the most important target for manufacturing companies. Many companies now design and manage their planning and control systems with a 'customer-oriented' focus. These customer-oriented manufacturing systems are required by manufacturing organizations that are driven by the requirements of their customers and the competitive pressures of the marketplace. In Figure 1.1, the development of manufacturing systems over the last century is represented. From this trend, it can be seen that CIM systems represent the leading edge of manufacturing technology at present. However, in many cases manufacturing is gradually becoming customer driven and is tending towards the extreme case of one-of-a-kind production (OKP) (Wortmann 1990). This one-off production environment is the extreme case of each individual product being completely customized to customer requirements.

Year	Manufacturing	
1920	Mass Production	Standardization
1970	Flexible Manufacturing Systems	↓
1980	CIM Systems	↓
2000?	One-of-a-kind Production	Customization

Figure 1.1 The trend in manufacturing systems.

This advent of customer-driven manufacturing has great consequences on current approaches to manufacturing planning and control philosophies. This book is not an attempt to describe an architecture for the extreme case of one-of-a-kind production, but instead it is an attempt to present an architecture and a series of ideas that can be applied to the many different trends and features encountered in the manufacturing environments of today.

In general, this book examines the topic of manufacturing planning and control (MPC) and its associated functions in the planning activities of an enterprise. In particular, the focus is on the strategic, tactical and operational planning and control activities of the enterprise and not on the automation or design of the manufacturing environment. As already stated, most manufacturing companies are moving away from making products for inventory purposes and they are becoming increasingly customer driven in their approach. This involves some rethinking, especially with regard to the MPC systems in use within the companies. To date much research has been devoted to the use of MRP II concepts as a tool that is mainly used for materials management in the factory, but not for the longer term planning functions (e.g. business planning, which is discussed in Chapter 3) and the detailed scheduling features that are needed for factories with both material and capacity constraints. This book attempts to revise the role of MRP II thinking in current manufacturing environments, and repositions it within a new framework. This new framework gives special emphasis to the integration of all levels of an MPC system and to the integration of the suppliers and customers in the total value chain of the enterprise.

One of the longer term planning functions that is not well addressed in many MRP II-type systems is termed master production scheduling or master planning. Vollmann *et al.* (1988) define the need for a study of master production scheduling as being of very great importance:

> ... the development of an effective master production scheduling (MPS) system is frequently cited as a critical element in obtaining the full benefits of MRP, and in managing effectively with MRP. The master schedule is generally acknowledged as the key input of MRP – the 'driver' of the MRP system.

The MPS function provides a mechanism for the dynamic interaction between the marketing/sales, manufacturing and engineering departments, and as such deserves special attention. A key argument in this book concerns the recognition of the MPS function as one of the most important activities in the production management systems (PMS) hierarchy. This topic is given special attention in Chapter 4.

Another important area that deserves special attention is the area of execution planning and control. The MRP II framework does not cater for this area in great enough detail and is unable to cope with the detailed planning and scheduling that is needed at the operational layer. Solutions for this level often include detailed mathematical algorithms and/or computer simulations in an attempt to find the 'best solutions' while optimizing both materials and capacity constraints. This topic is given special attention in Chapter 5.

There are three main themes in this book. These are as follows.

- Manufacturing environment. There is need for proper design and implementation of suitable manufacturing planning and control architectures that are specifically related to the production environment under consideration. Existing core MRP II techniques may be positioned within this architecture, but are not always a necessary component.

- Integration. The importance of the integration between the different levels in the manufacturing planning and control hierarchy needs to be emphasized. The information flows and the guidelines provided by the levels should be understood and utilized in their proper contexts.
- Different levels. The different levels within the manufacturing planning and control hierarchy are described in terms of business planning (strategic), master planning (tactical) and execution planning and control (operational level) in this book. This discussion is primarily motivated by the authors' involvement in the ESPRIT IMPACS project. Many of the ideas are based on user requirements gathered from several industry test sites and extensive user surveys.

In summary, the main issues to be addressed by this book may be loosely described by the following series of questions.

- What are the main elements involved in a manufacturing planning and control system?
- What are the main information flows between the different layers in a manufacturing planning and control system?
- What is the best way to achieve true integration between these different layers?
- What should a 'state-of-the-art' manufacturing planning and control system involve?
- What support tools can be developed to deal with the needs of manufacturing planning and control in the changing manufacturing environment of today?
- What should be the role of MRP-type approaches in manufacturing planning and control systems?

The remainder of this first chapter introduces some concepts behind manufacturing planning and control systems. The changing manufacturing environment is discussed as a starting point, and then a series of relevant concepts are introduced. Finally, a summary of existing practice in the domain is presented.

1.2 CHANGING MANUFACTURING ENVIRONMENT

Many manufacturing companies of today have to deal with requirements for:

- greatly reduced product life cycles;

- high product variety;
- unpredictable demand patterns;
- short customer lead times;
- large numbers of critical components with high lead times.

These requirements are really part of the challenge to companies to gain competitive advantage (Porter 1980). The challenge of gaining competitive advantage is making companies question their current approaches to production planning and control. Their task is also made more difficult because of the uncertainty of the market environment. This market uncertainty is illustrated in Figure 1.2, in which the complexity of the management of products and processes is shown on the vertical axis. This uncertainty–complexity grid was used for the first time by John Puttick of PA Consulting Group. Manufacturing organizations seek to decrease the uncertainty of the market and ease the complexity of management of the products and processes. The direction of the arrow shown in this figure reflects the desired direction in which companies would like to move (i.e. towards the lower right-hand corner). However, current market conditions are creating more specific requirements for products and less predictable demand patterns. Also, competitive pressures make companies respond with broader product ranges and complex systems and inevitably move towards the upper left-hand corner of Figure 1.2. The main challenges reside in being able to simplify and reduce the complexity so that things can remain 'manageable', and in attempting to shorten lead times. The easier it is to manage the company, then the higher the probability of being able to gain a competitive advantage.

The marketplace has changed dramatically in the last 20 or so years. It is important to realize that in the manufacturing environment of today the customer interaction with the manufacturer of the products is greater. Most companies are being forced to move away from making products in order to replenish the supply of products after they have been consumed by the customers. Customization to consumer needs is the emerging trend. Customers are no longer satisfied with standard products and increasingly consumers want a customized product that will uniquely fulfil their expectations. Previously, manufacturers produced standard components and stored them in a warehouse, which acted as a buffer for finished goods inventory. The customer then withdrew the products from the buffer, thereby having little or no interaction with the manufacturer. The factory scheduled the manufacture of products in order to keep the finished goods inventory at a specific level.

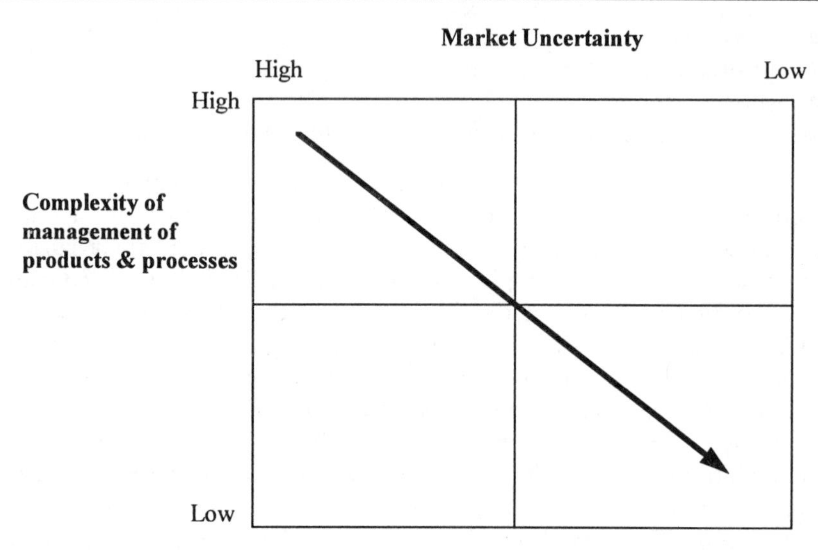

Figure 1.2 Market uncertainty versus complexity of management of products and processes.

In Figure 1.3A, the traditional type of system is depicted, in which customer and factory interaction was at a minimum. New pressures on manufacturing has led to a new situation, in which the customer interface with the factory is on a different basis to that which pertained previously. Typically there is no longer any finished goods inventory buffer from which the customer orders the products. Clients may now go directly to the manufacturer and express their preferences on product specifications and functionality. In effect, the customer is ordering a customized product, and this is illustrated in Figure 1.3B. In many cases there is only limited customization of products and standard components and subassemblies still exist.

The fact that there is a greater interaction between the customer and the manufacturer means that, in order to produce goods to customer specifications, the manufacturer must be in close contact with both the suppliers and the customers. The relationship with suppliers and customers is becoming one of the most important issues in the current manufacturing environment (Porter 1980). Customers request faster delivery times and are more specific in their needs for products. The MPS function often provides the means of closing the link between the manufacturer and the customer(s). In order to achieve an improved delivery performance, manufacturers need to have closer links with their suppliers to reduce material supply lead times. The approaches used in the area of supply chain management can help to close the link between the manufacturers and the supplier(s).

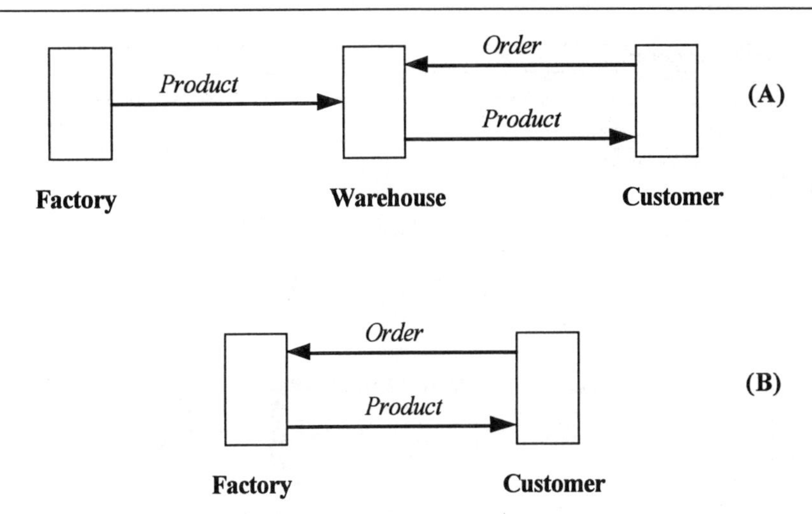

Figure 1.3 Customer–manufacturer interaction.

1.3 THE CHALLENGE OF INTEGRATION

Before continuing to describe the main themes of this book, it is first necessary to introduce some basic concepts that will be used frequently throughout the book. The first of these is the challenge of integration. This is presented in terms of integration from a functional viewpoint (e.g. different business processes and the associated information flows) and the perspective of the total **supply** or **value chain**.

1.3.1 INTEGRATED MANUFACTURING

A proposed overall model of an integrated manufacturing system is represented in Figure 1.4. This is a reference model that depicts, in a very global sense, the interactions that exist between various functional units within a CIM environment. Clearly this figure is greatly simplified but it gives a basic overall view. This proposed model of integrated manufacturing has been adapted from one developed initially at CIMRU, University College Galway, Ireland.

This model represents an integrated manufacturing environment from the perspective of information flows. The main objective in providing a truly integrated manufacturing system is to discover the nature of these flows, the time scales involved and the mechanism by which they operate. The model has some inherent weaknesses, however, as it does not represent the interactions between the various information flows, for example the linkage between the design and the customer order entry process.

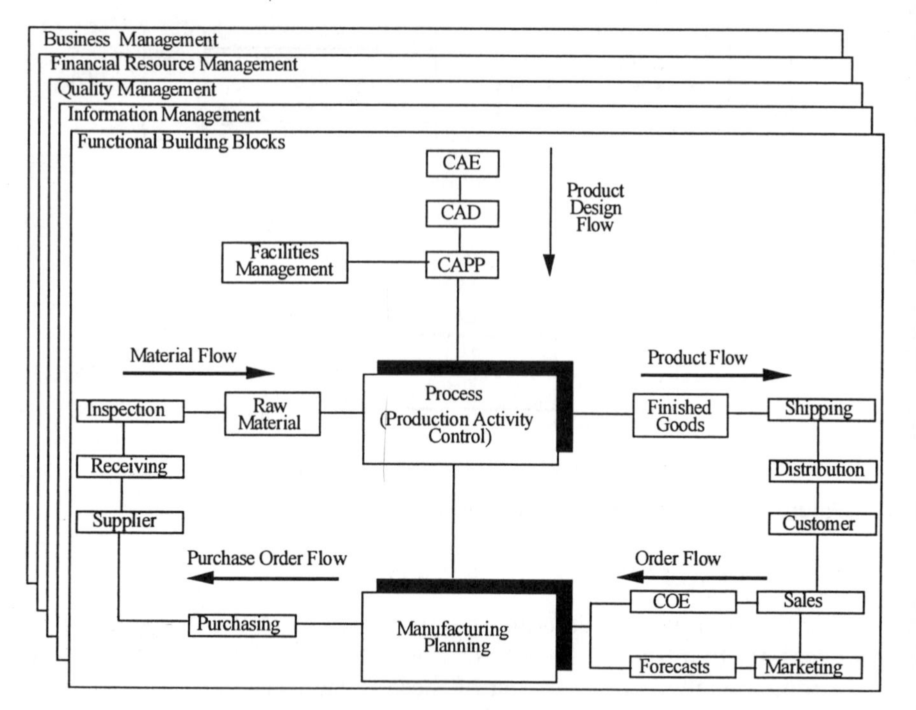

Figure 1.4 Proposed model for integrated manufacturing.

As shown in Figure 1.4, there are many 'layers' of management and issues that need to be dealt with in any organization. In very general terms these management layers consider the following aspects.

- The **business management** level deals with the strategic decisions that need to be made in order for companies to gain competitive advantage.
- The **financial resource management** level deals with the financial and accounting aspects of the business, attempting to ensure that targets are met and budgets are adhered to.
- The **quality management** level is an area of increasing importance in manufacturing today as customers are seeking products that work as specified and manufacturers are always attempting to get it 'right first time'.
- **Information management** is also becoming a major critical success factor in business today. Information technology and its associated aspects in business are revolutionizing the way that we work.

The information flows are depicted as being part of a functional building blocks layer in Figure 1.4. This is not an attempt at a software model or a detailed structured analysis and design of an ideal integrated manufacturing environment. It is instead a simplistic view of the main information flows and the basic building blocks or entities involved.

The **product design flow** is concerned with the conception of a design idea and the subsequent computer-aided engineering (CAE) and computer-aided design (CAD) that takes place in an organization. Indeed, this process may not involve the use of computers, but since we are talking about a CIM environment we assume that the terms CAE and CAD are generally applicable. Following on from CAE and CAD there is often a computer-aided process planning (CAPP) activity taking place in collaboration with a facilities management function. This involves the (re)design of the layout of the facilities in order to achieve the best means for production of any new product.

The **material flow** involves the information exchange between suppliers and the manufacturing process itself. This also involves a real flow of materials. The receiving department typically receives the goods and associated information from the suppliers. In many cases this is then followed by an inspection process before materials actually get placed into stock for use on the shop floor.

The **purchase order flow** is the starting point for the ordering of materials from suppliers. These purchase orders are generated by the production management systems in the manufacturing planning department. This is typically done through the purchasing department.

The **order flow** of information comes from the customers. This is typically received by the sales and marketing departments, and in some cases may actually be forecasts and orders. There is typically a formal customer order entry (COE) system that processes orders before they are handed over to the manufacturing planning department. This may involve regular meetings with manufacturing and sales/marketing personnel in order to give realistic promise dates to the customers.

The **product flow** is the vital flow in the organization. This involves the completion of the manufacturing on the shop floor and the addition to finished goods inventory, or, ideally, the products go directly to the customer via shipping and distribution planning departments.

It is commonly believed that future production systems need to be more integrated than those that exist today. Unfortunately, there is not enough agreement about the meaning of the term 'integration'. Some authors have described integration as 'interacting business processes of product and product design, manufacturing planning and control, and production processes . . . all linked via information technology' (Falster *et al.* 1991). Throughout the remainder of this book, we attempt to put forward some useful ideas and concepts that may aid in the design and successful operation of an integrated MPC system. This integrated MPC system should be seen as a subset of the overall desired integration needed within manufacturing organizations.

Traditionally, CIM has been discussed mainly in terms of the 'four walls' of the manufacturing process. However, in the competitive manufacturing environments of today, much more emphasis is being placed on closer interaction with the suppliers (e.g. supply chain management) and the customers (e.g. distribution management). This means that more emphasis is also being placed on the extreme left- and right-hand sides of the basic model presented in Figure 1.4. The complete cycle from supplier through to end customer is often discussed in terms of the supply or value chain.

1.3.2 SUPPLY OR VALUE CHAIN

The supply or value chain is a concept introduced by Porter (1985) in order to diagnose and enhance competitive advantage. Porter states 'Value chain analysis helps a manager to separate the underlying activities a firm performs in designing, producing, marketing and distributing its product or service.'

Other notable quotes from the work of Porter also emphasize the usefulness of value chain analysis. For example:

> To diagnose competitive advantage, it is necessary to define a firm's value chain for competing in a particular industry. Starting with the generic chain, individual value activities are identified in the particular firm . . .
>
> Although value activities are the building blocks of competitive advantage, the value chain is not a collection of independent activities but a system of interdependent activities. Value activities are related by linkages within the value chain. Linkages are relationships between the way one value activity is performed and the cost or performance of another . . .
>
> Linkages can lead to competitive advantage in two ways: optimization and coordination. Linkages often reflect trade-offs among activities to achieve the same overall result. For example, a more costly product design, more stringent materials specifications, or greater in-process inspection may reduce service costs.

The value chain is a useful model to express the business and organization structures now emerging. To demonstrate some changes in the value chain, consider Figure 1.5, which has been developed by Browne *et al.* (1994).

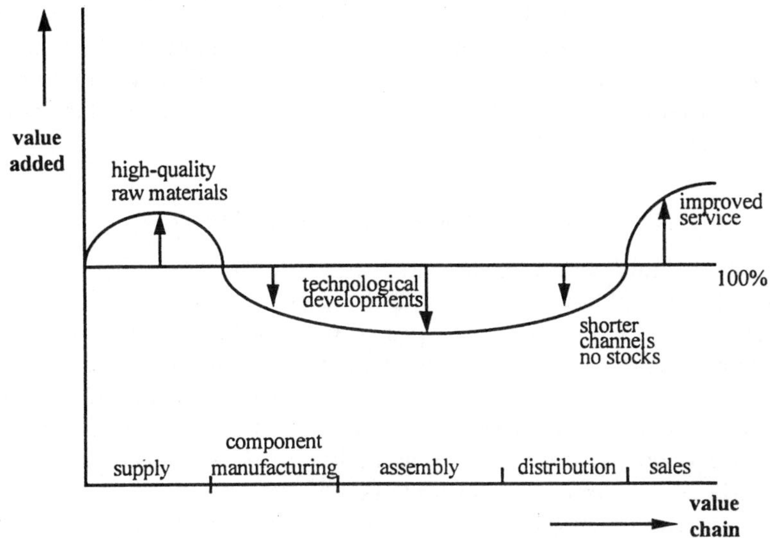

Figure 1.5 Changes in the value chain (Browne *et al.* 1994).

Suppose that the contribution of each element to the added value of the final product is normalized to 100%. In this context, the added value refers to any addition of materials or utilization of resources/skills that lead to an increase in the value of the product to the customer. Then Figure 1.5 suggests five effects as follows.

- New materials such as highly alloyed steel, composite materials, etc., will probably lead to an increase in value added in material supply.
- Component manufacturing will probably move towards lower value added, because of the effects of miniaturization and other manufacturing technologies.
- Assembly is likely to move towards lower added value also, because much functionality is already available in the components.
- Physical distribution will probably involve lower costs if the interaction between the customer and the manufacturer becomes more intensive.
- Improved service around the product will increase the added value in sales and after-sales activities.

These or similar effects underline the importance of the need to analyse the value chain of manufacturing organizations. The manufacturing system must now be seen in the context of the value chain. This is often termed the extended enterprise. If the challenge was to realize integration for the individual factory, the challenge now is to facilitate inter-enterprise integration across the value chain. This topic will be further explored in Chapter 2. In the next section, an attempt is made to provide a categorization of the different types of manufacturing environments.

1.4 DIFFERENT MANUFACTURING ENVIRONMENTS

In order to understand clearly the need for a new approach to manufacturing planning and control, it is necessary to have an understanding of the different types of manufacturing environments. Different analyses of current manufacturing systems have resulted in different typologies of the systems (Wemmerlov 1984). The following typology describes four classic types of manufacturing environments. These are **make to stock (MTS), assemble to order (ATO), make to order (MTO)** and **engineer to order (ETO)**.

Make to stock characterizes the manufacture of products based on a well-known and relatively predictable demand mix. MTS companies try to anticipate product demand and produce to stock so that products are available to meet product demand. Products are typically stored in a central finished goods warehouse. They may also be distributed halfway along the distribution network or distributed to the final outlets of the distribution network in anticipation of actual demand. MTS companies are characterized by a limited number of products, produced and possibly distributed based upon forecasts. Forecasts are usually developed by product family or even product, which is possible in view of the mostly sufficiently high sales volumes. In this environment interaction with the customer is rather distant, the production volume of each sales unit is high, and customer delivery time, determined by the availability of finished goods inventory, is relatively short (ideally zero lead time between product order and delivery). The finished goods act as a buffer against uncertain demand and stockouts, which usually result in lost sales. A typical MTS system is portrayed in Figure 1.3A. The MTS system has the advantage of normally having quick delivery time, but inventory costs are high and customers are unable to express preferences as to the product design. The MTS environment is characterized by reasonably long and predictable product life cycles.

Assemble to order involves having the same core assemblies for most products and the ability to vary all other components of the final assembly. ATO companies have a hybrid planning and control approach and are often forced, by intense competition, to provide a wide range of products with short customer delivery lead time. The hybrid planning and control approach consists in modularizing products into a series of semifinished modules, managing the module manufacturing stage in an MTS fashion, while managing the final assembly manufacturing stage in an MTO fashion. In effect, ATO companies attempt to manufacture a series of semifinished modules that can be quickly assembled together to form a final product requirement of a customer. Products are designed in such a way that a wide range of products can be assembled from a more limited range of modules. Demands for modules can therefore more easily be forecasted, enabling an MTS approach. Final assembly is driven by firm demands only. In a manufacturing environment working to this strategy, contact with the customer occurs primarily only at a sales level. The delivery time is of medium length and is based on the availability of major subassemblies. ATO companies handle demand uncertainty by overplanning components and subassemblies. Assembly only takes place on receipt of an order, and buffers of modules or options may exist. The product routeing in the factory is typically fixed. No final inventory buffer exists and the customer has limited input into the design of the product. ATO companies attempt to provide competitiveness in terms of both lead time and product range.

Make to order involves having all the components available along with the engineering designs, but the product is not actually specified. The finished product from this system is partially one of a kind, but not entirely one of a kind because the final product is not usually designed from a basic specification. Manufacturing of the product begins with receipt of an order, and the configuration of the product is likely to change from the initial specification during the course of processing. Products are mostly customized to the specific needs of each individual customer, and the bill of material (BoM) is usually unique for each product. The BoM has been defined by the American Production and Inventory Control Society (APICS) (1980) as 'A listing of all the subassemblies, parts and raw materials that go into a parent assembly showing the quantity of each required to make an assembly'. As a result, the product range is very broad with very limited sales by individual product. Interaction with the client is extensive and is based on sales and engineering information. The MTO activities are characterized by a low degree of uncertainty, yet the customer lead time ranges from medium to long. However, this lead time can be reduced by purchasing materials based upon forecast. Sales volumes are very difficult to forecast. Therefore, procurement to forecast should be limited to long lead time items with high usage in many product variants and low probability of obsolescence. Promise dates for completion of orders are mainly based on the available capacity in manufacturing and engineering.

Finally, engineer to order is an extension of the MTO system with the engineering design of the product being almost totally based on customer specifications. The same characteristics apply here as to the case of MTO, but customer interaction is even greater. True one-of-a-kind products are engineered to order.

Some of main differences between 'to stock' and 'to order' manufacturing are summarized in Table 1.1.

Table 1.1 Differences between 'to stock' and 'to order' manufacturing

	To stock	To order
Production	To forecast or stocking level	To order
Products	Predefined	Customer defined
Part numbers	Preidentified	Unique identity
Bill of routeing	Pre-engineered	Created at order entry

As we move from MTS to ETO environments the 'customer order decoupling point' defines the point after which any material is dedicated to a particular customer order. The positioning of this customer order decoupling point is vital, as it defines the parts of the process that are driven by customer orders and the parts that are driven by forecasts (Van Veen 1990). The positioning of the customer order decoupling point in the different manufacturing environments is illustrated in Figure 1.6. In general, the decoupling point between the manufacturing-driven and sales-driven operations is tending to move towards earlier stages of the manufacturing process. The customer order decoupling point concept will be discussed more thoroughly in Chapter 4.

Wemmerlov (1984) states that a manufacturing company's operation can be classified as MTS, ATO, MTO or a combination of these. Deciding which approach a company should adopt is a strategic decision, and will strongly affect the way that a company conducts its manufacturing planning and control activities. MTS and MTO represent two 'pure' manufacturing strategies, while ATO is a hybrid strategy. It is likely that most companies originate as either MTS or MTO firms and later progress into the ATO stage (Figure 1.7).

A company starting out as an MTO firm may choose to get into ATO manufacturing because of an expanding volume or a strong similarity between some of its products. The move to ATO is made in order to capitalize on an increased demand and the possibility of reducing customer delivery times for a subset of its products. Alternatively, an ATO firm may previously have produced to stock. This move to ATO is usually made to obtain a greater market share through offering a larger variety and better service. Tables 1.2–1.4 compare and contrast the different approaches. These tables are based on the paper by Wemmerlov (1984) on ATO manufacturing.

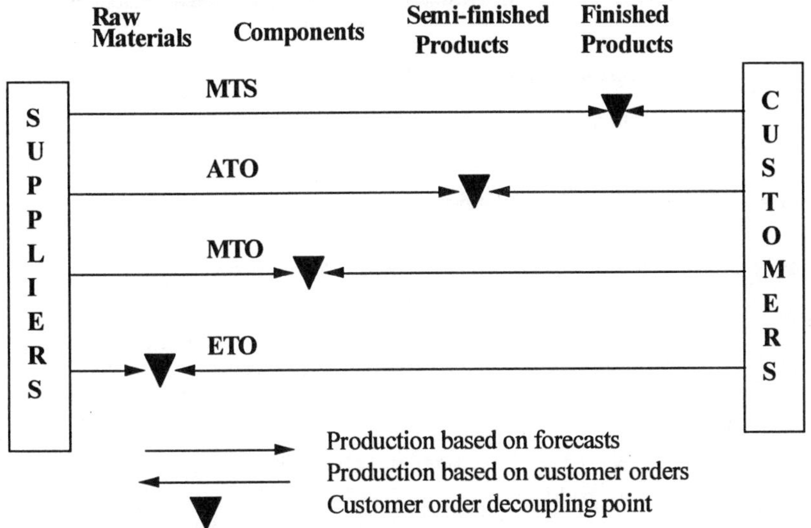

Figure 1.6 Customer order decoupling point.

The master planning approach used in companies is often directly related to the type of manufacturing environment. Each individual industry segment has its own specific requirements, and there is a strong relationship between the positioning of a certain company and its requirements with respect to production planning and control systems and techniques. Master planning is of particular importance as it is positioned at the interface between the production system and its market and therefore should closely reflect the nature of both.

In an MTS environment, the master planning activity typically has the finished product as the MPS item. However, in the case of long procurement and production lead times, master planning may be carried out differently beyond a certain time fence. For example, groups of products or components may be master scheduled directly. This is allowable as the plans beyond a certain time fence do not directly affect the final production steps in the finalization of a product. Therefore, in MTS companies the MPS is very closely related to the final assembly schedule.

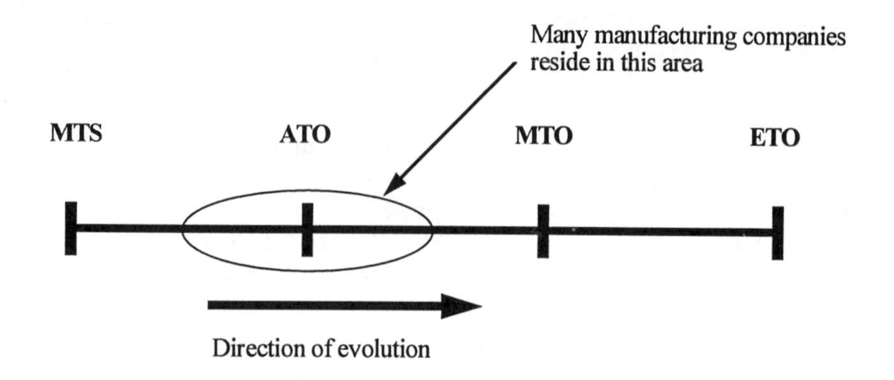

Figure 1.7 Evolution of manufacturing.

Table 1.2 Customer interaction in different manufacturing environments

Aspect	MTS	ATO	MTO	ETO
Interface between manufacturing and customer	Low/Distant	Primarily at sales level	Engineering and sales level	Primarily at engineering level
Delivery time	Short	Medium	Variable	Variable*
Production volume of each sales unit	High	Medium	Low	Very low
Product range	Medium	High	Low	Very low
Order promising (based on)	Available finished goods	Availability of components and major subassemblies	Capacity for manufacturing engineering	Capacity for manufacturing engineering

*MTO and ETO environments normally have long delivery times.

Table 1.3 Production planning in different manufacturing environments

Aspect	MTS	ATO	MTO	ETO
Basis for production planning and scheduling	Forecast	Forecast and backlog	Backlog and orders	Customer orders
Handling of demand uncertainty	Safety stocks of sales units	Overplanning of components and subassemblies	Little uncertainty exists	No control

Table 1.4 Implications for master planning in different manufacturing environments

Aspect	MTS	ATO	MTO	ETO
MPS unit	Sales unit	Major components	End products	End products
Final assembly scheduling*	Close correspondence to the master schedule	Determined by customer orders received by order entry	Covers most of the assembly operations	Covers all of the assembly operations
Bill of material structuring	Standard BoMs (one BoM for each sales item)	Planning BoMs† are used	BoMs are unique and created for each customer order	BoMs are unique and created for each customer order

* The term final assembly schedule has been defined by Vollmann *et al.* (1988): 'a statement of the exact set of end item products to be built over some time period, the final assembly schedule controls that portion of the business from fabricated components to shipable products'.

† Planning BoMs have been defined by APICS (1980) as 'an artificial grouping of items, in bill of material format used to facilitate master scheduling and/or material planning'.

In an ATO environment, the master planning activity typically takes place at the level of standard or core modules. The master planning unit may also vary beyond a certain time fence similar to the MTS environment. Assemble-to-order firms have a very complex problem in defining an MPS unit. The ATO business is typified by an almost limitless number of possible end item configurations, all of which are made from a combination of basic components and subassemblies. Customer lead times are typically shorter than total lead times, so production must be started in anticipation of customer orders. The large number of end item possibilities makes forecasting of exact end items extremely difficult. The ATO environment does not normally master schedule end items. The MPS unit is usually stated in terms of the-standard and core modules. Planning bills of material may be used to identify sets of common parts and options. The planning bill of material represents an artificial end product which cannot be built. It exists purely for master scheduling purposes, and it relies on the final assembly schedule to make a saleable product.

The MTO company generally carries no finished goods inventory and builds each customer order as needed. This form of production is often necessary where there are very large numbers of possible product configurations, and therefore great difficulty in anticipating the likely precise needs of the customer. Here, the MPS unit is typically defined as the particular end item or set of items that constitutes a customer order. The definition is difficult, since part of the job is to define the product, that is design. Production regularly commences before a complete product definition and bill of materials have been determined. In this case procurement is done against a forecast and master planning may take place at the component level.

The ETO company generally has the same features as the MTO companies, with main differences being that it does not plan in terms of forecast and master planning is done against available capacity.

Traditionally, manufacturing planning and control theory has mainly been associated with product-oriented companies. Their focus is typically on streamlined production methods, minimization of overheads, etc., as well as high-quality products in the marketplace. However, in the changing manufacturing arena of today, many companies are finding it harder to compete. For example, the computer industry has changed rapidly in the last 20 years. Many smaller companies now have the capability to produce cheap high-quality computers. Companies are finding that they have to diversify and have begun selling their services, knowledge, computer solutions, etc. This is also caused to some extent by the changing costs involved in labour overheads. Companies often find it much cheaper to manufacture their products in other countries. Therefore, it could be argued that companies should consider having both product and capability orientations in order to survive in highly competitive market segments.

In Figure 1.8, the uncertainty–complexity grid is shown once again. This time it includes the types of goods that are typically manufactured in each quadrant.

The **capital goods** industry is typified by having single or few customers, manufacturing complex or customized products, and a relatively simple distribution system due mainly to low volumes of products. Production is typically of the MTO type owing to the high uncertainty of the market. Production planning often requires good project planning at a higher level, and usually bills of material are quite complex. This gives rise to a difficult materials management problem. Therefore, MRP II systems would be suitable for this type of industry and this is actually the industry from which MRP II thinking originated. The important competencies required of the capital goods industry include customer order management (in order to facilitate customized requirements and design inputs), two-level planning (a combination of project planning at the higher level and MRP at the lower level) and control, flexible production processes and subcontracting capabilities (especially in cases where in-house capacity or capability may be in short supply).

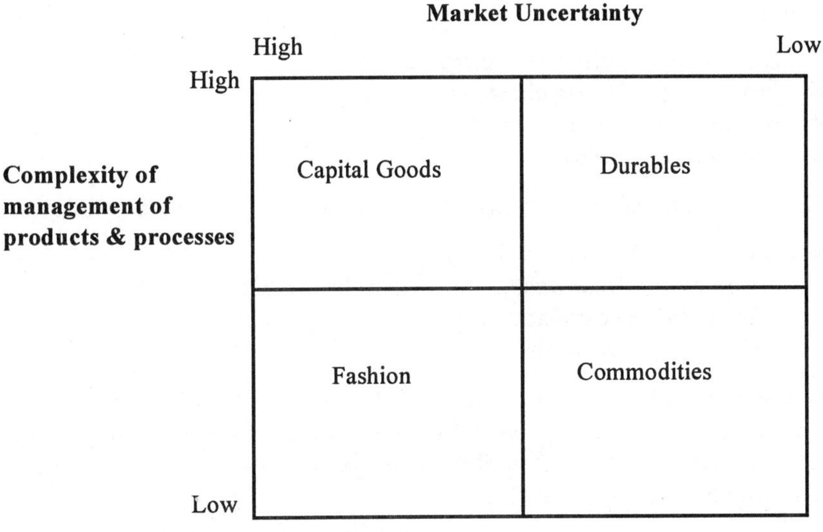

Figure 1.8 Type of goods manufactured.

The key characteristics of the durables industry include a limited range of products and a high volume of customer orders. The products are typically highly complex, but with a very certain market. Product differentiation is offered mainly as a means to gain market share and remain competitive. Costs are often controlled by producing a standard range of modules from which a wide range of possible products can be produced. Production is thus typically of the ATO type, and entails complex customer order promising. The required MPC competencies include customer order promising (linked with customer order configuration), and master planning executed at the intermediate level of the bills of material.

Companies in the **commodities** industry typically have many customers, make standard products in high volumes and have few major suppliers. Production is typically of the MTS type owing to the low uncertainty of the market. The industry can be associated with the manufacture of 'bulk' goods. The required MPS competencies include sales order processing, and good distribution managements.

The key characteristics of the **fashion** industry are many customers with close links to the factory and a variable supply base. Companies in this industry require the ability to respond rapidly and must have a very flexible manufacturing environment. In general, the required competency is flexibility.

In summary, the **critical success factors** for each type are suggested as follows:

- capital goods: fitness for purpose;
- durables: providing choice cost-effectively;
- commodities: price;
- fashion: availability.

Finally, the critical competencies for each type are suggested as follows:

- capital goods: effective design;
- durables: modular design, flow manufacturing;
- commodities: cost leadership;
- fashion: rapid response.

In the remainder of this chapter, a brief review of an industry survey carried as part of the project that motivated this work is described. Chapter 2 is devoted to describing some basic streams in manufacturing planning and control thinking.

1.5 REVIEW OF INDUSTRY PRACTICE AND LITERATURE

IMPACS was the name given to a project sponsored from 1989 to 1992 by the European Commission under the umbrella of ESPRIT, the European Strategic Programme for Research in Information Technology. IMPACS is an acronym for integrated manufacturing planning and control systems, and the project aimed at evaluating existing concepts for production planning and control and making recommendations for improving these concepts. The consortium involved the following companies/institutions: Alcatel (France and Belgium), Digital Equipment Corporation (Ireland), PA Consulting Group (UK and Belgium), Comau (Italy), Centunion (Spain), the CIM Research Unit at University College Galway (Ireland) and the GRAI Laboratory at the University of Bordeaux (France).

The IMPACS project team designed and developed decision support systems for business planning, master production scheduling and factory coordination issues in the area of manufacturing planning and control. These layers respectively correspond closely to the strategic, tactical and execution layers of a manufacturing planning and control architecture (Chapter 2). In the first year of the IMPACS project, each level of the hierarchy was examined in detail and functional models and software prototypes were developed. During this time, the IMPACS project team carried out an extensive survey of the state of the art in the domain. This survey provided an insight into the problems with existing systems. In summary, the systems focused more on communications and data processing rather than on the organizational and management domains. During the second year, a collection of user requirements were gathered from four test sites, and testing began on implementations of prototype software. During the final year of the project the focus was on the development of a fully integrated prototype solution, linking the different layers of the planning and control hierarchy. As well as working on the inherent functions of business planning, master production scheduling and factory coordination, there was also research undertaken in the architecture and information technology domains. These were effectively sanity checks on the work, ensuring integration between the different layers by using advanced modelling techniques (architecture) and state-of-the-art software development solutions (information technology). The following is a summary of the more specific achievements of the IMPACS project (IMPACS 1992).

1.5.1 ARCHITECTURE

This element of the project involved the development of a modelling framework, a complete and integrated conceptual reference architecture and a methodology to implement the IMPACS architecture. A decision support system for modelling the architecture, entitled CAGIM (computer-aided GRAI IMPACS Method), was also developed as a support for this work. This system is still under development and has been used in a number of projects by the GRAI Laboratory, Bordeaux.

1.5.2 BUSINESS PLANNING

A framework for manufacturing business planning was developed. This framework identified the main decision categories involved in the development of a manufacturing strategy and the formulation of a long-range production plan. Both qualitative (highly unstructured information) and quantitative (detailed 'numerical' information) software prototypes were developed. One of these was an expert system-based rules base for qualitative modelling to aid the **make or buy** decision. Another was a tool integrating financial, manufacturing and sourcing issues in order to aid the make or buy and capacity decisions within a company. The latter tool was implemented successfully within a test site (Digital Equipment Corporation, Galway). The business planning topic will be discussed in detail in Chapter 3.

1.5.3 MASTER PRODUCTION SCHEDULING

This area was identified as the central focus in the planning and control hierarchy. It provided the link between the long-term strategy formulation and the execution and purchasing planning at the lower levels. The main software result in this area was an MPS planning table system that had many innovative features, such as customer order entry linkages to the MPS, planning bills of material, integration with long-range plans and forecasts and resource analysis. This area will be covered in detail in Chapter 4.

1.5.4 FACTORY COORDINATION

The main result in this domain was the development and implementation of a factory coordination scheduler in a test site (Alcatel BELL, Hoboken, Belgium). The scheduler creates manufacturing orders for a set of planned orders received from the companies' requirements planning system. These manufacturing orders are used to decouple operational execution from tactical planning. The manufacturing orders are broken down into a series of orders for each of the different 'cells' in the factory using routeing and process information. These cell orders can be scheduled using various strategies and the results compared to select the best solution. Capacity profiles are also provided to enhance the schedule validation process. This topic will be discussed in Chapter 5.

1.5.5 INFORMATION TECHNOLOGY

The information technology element of the IMPACS work initially provided a state-of-the-art report on relevant tools and techniques in the area of manufacturing planning and control. This later provided the input for an IMPACS information technology model for the development of group decision support systems. Quality assurance guidelines and hardware/software portability issues were also addressed in the development of all the software for the project.

In the first part of this chapter, the need for a recognition of the fact that different manufacturing environments have different requirements in terms of their manufacturing planning and control systems was broadly described. The need for proper integration between the different levels in a manufacturing planning and control hierarchy was also described. In the remainder of this book, the authors will attempt to illustrate a framework that can be used to help solve issues of this kind. In general the authors will attempt to:

- demonstrate that different approaches are needed in different industry segments today and illustrate that certain approaches, such as MRP, may have limited applicability; standard tools such as MRP may be included in an overall production planning and control architecture, which should first be tailored to the industry-specific requirements;
- propose new ideas to improve the planning activities at the strategic, tactical and execution layers in a manufacturing organization.

Before continuing to specify the details of this framework, it is necessary to give a review of related research and industry practice in this area. In the following, a synthesis of the ideas gathered during a literature survey and an industrial survey (interviews and questionnaires) are presented. The work formed the basis for the development of software prototypes within the IMPACS project. These prototypes were designed based on the requirements of several industrial users. One of these prototypes will be described in Chapter 3. This chapter only contains brief summaries. These issues will be elaborated upon in the relevant chapters later in this book.

1.5.6 STRATEGIC PLANNING PRACTICE

In all cases surveyed, there was a considerable amount of difference in the understanding and approach towards strategic planning. Attention was mainly concentrated on issues relating to quality, lead times, low costs and organizational structures. The time scale for strategic planning activities was dependent on the time scale of the business strategy, the lead time for the addition of capacity and product life cycles. The most popular horizon was 5 years. The only 'tools' used in the creation of the strategy plans were spreadsheets on personal computers to enable rapid modelling of volumes and financial implications of shifts in product volume and mix, headcount, etc.

1.5.7 TACTICAL PLANNING PRACTICE

The review of the tactical planning practice highlighted numerous specific capabilities that were missing in current planning/scheduling systems. These included simulation capabilities, on-line MPS updates, pegging facilities, performance measurement, planning bills of material, easy-to-use graphical support tools, etc. Many of these topics are quite detailed in nature and are addressed in Chapter 4.

1.5.8 EXECUTION PLANNING PRACTICE

The execution planning practice review consisted mainly in questions in the area of factory coordination (planning and managing the flow of orders throughout a series of individual manufacturing cells) and not detailed shop floor scheduling (at the level of individual resources within each manufacturing cell). This factory coordination level of planning typically takes place after an MRP-type calculation process has taken place and the factory planners are in a position to launch work orders on the shop floor. The main considerations arising from the survey included the need for more effective production environment design, identification of critical resources, subcontracting, handling of exceptions, etc. In general, there was a clear need for better factory planning and scheduling procedures and tools. This included more advanced synchronization between different cells within factories as well as the synchronization with the upper level planning and customer order entry procedures. This topic will be discussed in Chapter 5.

The state-of-the-art review helped identify many requirements for manufacturing planning and control systems. These requirements are closely linked with the overall themes of this book, as they highlight the need to take a different view of manufacturing planning and control to that currently existing in traditional MRP II thinking. The integration between the different levels of the manufacturing planning and control hierarchy have been neglected in many places and little attention has been given to the study of tactical planning or scheduling in hybrid manufacturing environments. In conclusion, it is suggested that there is a need to review the existing approaches to manufacturing planning and control. This book is an attempt at describing a framework within which companies may base their manufacturing planning and control approach. Some companies have environments in which MRP II thinking is highly suitable and others do not need it at all. However, if MRP II is selected it should be recognized that master production scheduling is the key to its success. Some basic streams involved in manufacturing planning and control thinking are presented in the next chapter.

Basic streams in manufacturing planning and control thinking

2

2.1 INTRODUCTION

In this chapter some of the concepts and ideas involved in manufacturing planning and control thinking are described. This should be useful as an introduction to some of the background to the material presented in the remaining chapters of this book. The remainder of this chapter is structured as follows.

- The MRP II framework is presented in terms of strategic, tactical and operational/execution layers.
- **Just-in-time** and **synchronous manufacturing** are briefly described as the other two most well-known approaches.
- Some drawbacks to the MRP II approach are presented.
- A revised manufacturing planning and control architecture is discussed which lays the foundation for detailed discussions given in Chapters 3–5.
- A discussion on an expanded future role for total **enterprise planning** is introduced.

2.2 THE MRP II FRAMEWORK

During recent years, numerous projects have addressed the development and implementation of **integrated** manufacturing planning and control systems. The IMPACS project worked on this area for discrete manufacturing environments. IMPACS was based on the premise that the MRP II approach (Orlicky 1974, Wight 1981) needs improvement, and attention should be given to the different layers of the manufacturing planning and control systems hierarchy (Higgins and Browne 1992). An outline sketch of this manufacturing planning and control systems hierarchy is shown in Figure 2.1. The architecture depicted in this figure extends from the strategic to the operational (or execution) level. These levels represent different planning horizons. The length of the time horizons depend on the production environment under consideration. The strategic planning horizon may cover 1 to 5 years, tactical planning 1 month to 1 year and execution planning real time to 1 week.

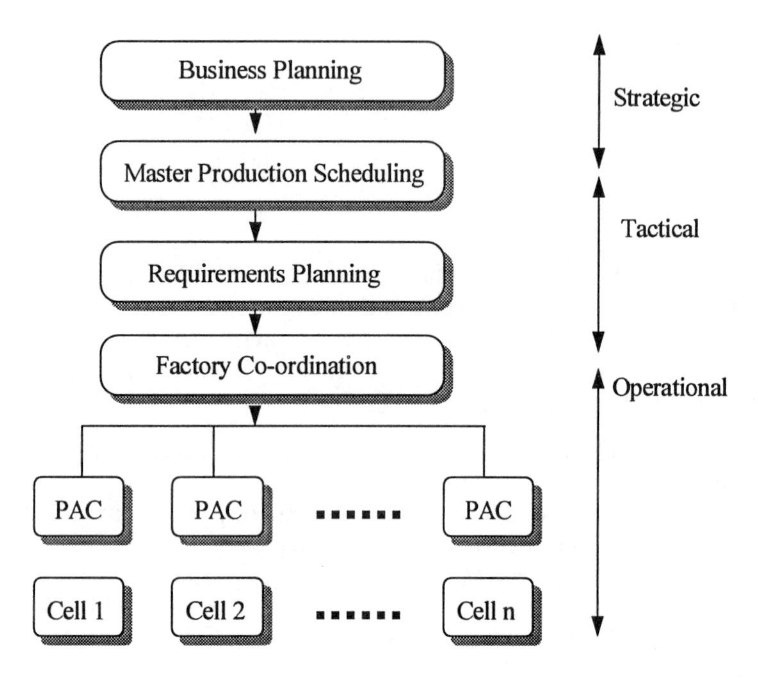

Figure 2.1 Manufacturing planning and control systems architecture.

This architecture reflects a situation in which a factory has been decomposed as far as possible into a series of group technology-based production cells (Burbidge 1989), where each cell is responsible for a family of its products, components or processes and each cell is controlled by a production activity control (PAC) system. Another possibility is that each group is actually geographically dispersed, that is there are a number of different focused factories. The factory coordination layer ensures that the individual cells/factories interact to meet an overall production plan. This overall production plan relates to the work orders that need to be scheduled and executed in the cells/factories. The work orders are input from a requirements planning layer that in most cases includes an MRP processor. The requirements planning layer receives it main orders in terms of overall products or assemblies from the master production scheduling level, which in turn is a quantitative description of guidelines sent down by the business planning level.

In our view the MRP approach needs to be understood more clearly in terms of the manufacturing environment under consideration. In the remainder of this section some of the basic concepts underlying the MRP approach are described.

2.2.1 MRP CONCEPTS

MRP thinking has revolutionized manufacturing planning and control since J. Orlicky wrote his book in 1974. MRP systems aim at planning and controlling purchasing and manufacturing operations in response to a certain manufacturing programme. The MRP logic is entirely based upon the concept of dependent demands.

In a manufacturing environment, material conversion stages determine the relationships between raw materials, purchase parts, component parts, subassemblies and finished products. Because of these relationships, demands for a given item are dependent upon what is scheduled to be produced for the next conversion stage. MRP expresses the relationships between parent and child items by means of bills of material.

The MRP logic is an approach for planning and controlling all of the items, in response to a certain manufacturing programme, and according to the concept of dependent demand. The logic can be summarized as an iteration of three consecutive steps [see Browne *et al.* (1988) for a detailed example of the MRP calculation process], as follows:

- **netting** against available inventory;
- **calculation of planned orders**;
- **bill of material explosion** to calculate gross requirements for dependent items.

However, the MRP process needs an input, namely the overall manufacturing programme, which is often called an MPS. The basic structure of an MRP explosion process is depicted in Figure 2.2. The details behind the logic of this method are described in numerous books and publications. The best known of these was written by one of the initial developers of the technique, Joseph Orlicky (1974).

The main objectives of MRP are to determine **what** and **how much** to order (both purchasing and manufacturing orders). Before the widespread use of MRP, the planning of manufacturing inventory and production was generally handled through inventory control approaches. The implicit assumption of these approaches is that the replenishment of particular inventory items can be planned independently of each other. In developing the ideas of MRP, Orlicky offered several important insights into inventory management that have revolutionized inventory management practice. These can be summarized as follows.

- Manufacturing inventory, unlike finished goods or service parts inventory, cannot usefully be treated as independent items. The demand for component items is dependent on the demand for the assemblies of which they are a part.
- Once a time-phased schedule of requirements for top-level assemblies is put in place (MPS), it follows that the dependent time-phased requirements for all components can be calculated.
- The assumptions underlying inventory control models usually involve a uniform or at least well-defined demand pattern. However, the dependency of components demands on the demand for their parent items gives rise to a phenomenon of discontinuous demand at the component level.
- A computer system can provide the basic processing capability to perform the necessary calculations.

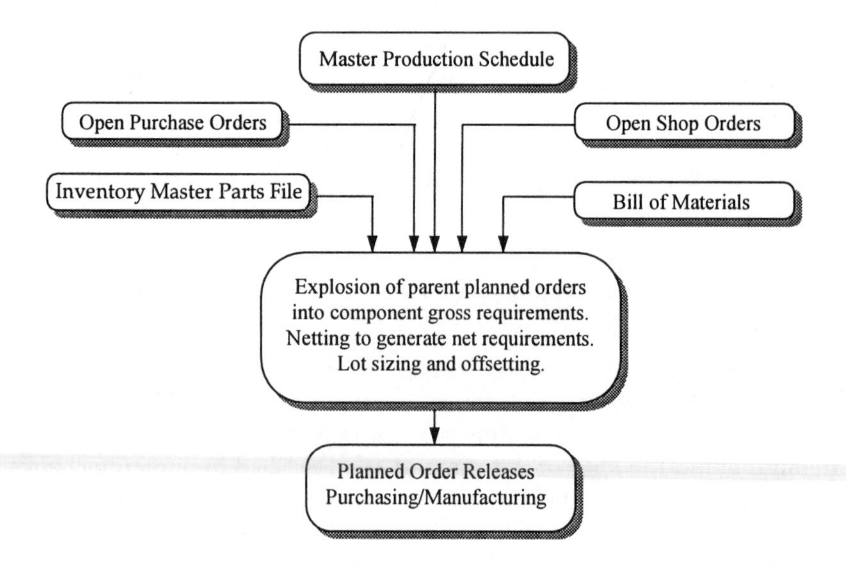

Figure 2.2 Basic structure of MRP systems.

MRP can thus be seen as discrete control on the flow of materials rather than control of the level of inventory. MRP orders the components that are required to maintain manufacturing flow. As shown in Figure 2.2, the MRP calculation engine is driven by the MPS, production structure information (bills of materials), current inventory records and component lead time information.

There are two different approaches to the use of MRP: **regenerative** and **net change**. Regenerative MRP refers to the case when the whole MRP explosion process is repeated on a regular basis. The net change approach involves the explosion of only the alterations to the inputs taken into account when compared with the previous MRP explosion.

Manufacturing resource planning (MRP II) is an extension of MRP features to support many other manufacturing functions beyond material planning, inventory control and bill of materials control. MRP II evolved from MRP through a gradual series of extensions to MRP functionality. The basic structure of MRP is depicted in Figure 2.3.

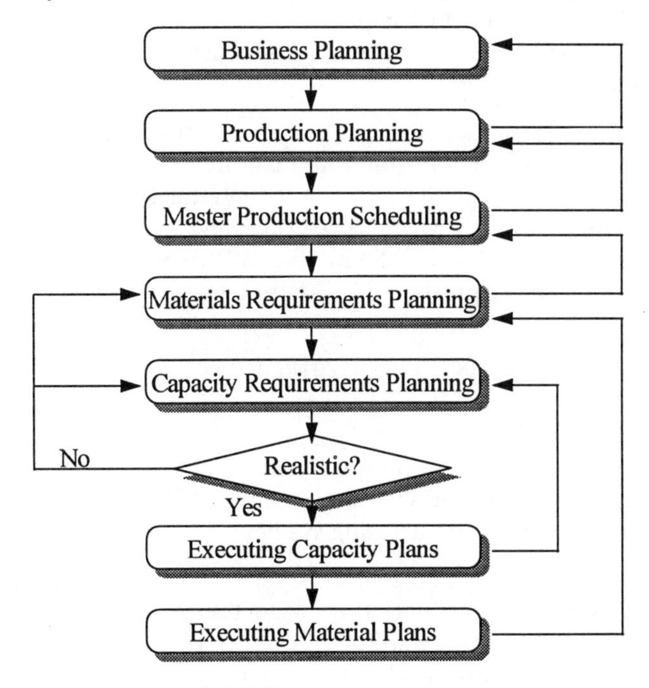

Figure 2.3 Basic structure of MRP II.

MRP II (manufacturing resource planning) has been defined by APICS as:

> A method for the effective planning of all the resources of the manufacturing company. Ideally it addresses operational planning in units, financial planning in dollars, and has a simulation capability to answer 'what-if' questions. It is made up of a variety of functions, each linked together: business planning, production planning, master production scheduling, material requirements planning, capacity requirements planning and the execution systems for capacity and priority. Outputs from these systems would be integrated with financial reports, such as the business plan, purchase commitment report, shipping budget, inventory production in dollars, etc. Manufacturing resource planning is a direct outgrowth and extension of MRP. Often referred to as MRP II (cf. closed-loop MRP).

Wight (1981) summarizes the characteristics of MRP II as follows.

- The operating and financial system are one and the same.
- MRP II contains 'what if?' capabilities.
- MRP II involves every facet of the business from planning to execution.

Essentially, MRP II can be described as 'closed-loop' MRP with connections to business and financial planning, reporting facilities and 'what if?' simulation capabilities. Today, the terms MRP II and MRP are often collectively termed MRP. Three different layers of planning and control can be distinguished in the MRP production planning architecture. These are presented in Figure 2.4.

2.2.2 STRATEGIC PLANNING AND CONTROL

Strategic manufacturing issues relate to:

- the determination of the products to be manufactured;
- the matching of products to markets and customers' expectations;
- the design of the manufacturing system.

Strategic manufacturing aims to ensure short production lead times and sufficient flexibility to facilitate the production of the required variety and mix of products for the market.

The strategic planning and control layer consists of the sales and operations planning activity. Sales and operations planning is often called production planning. The authors, however, believe that the latter term often causes confusion, whereas the first term gives a good description of what the activity is actually about.

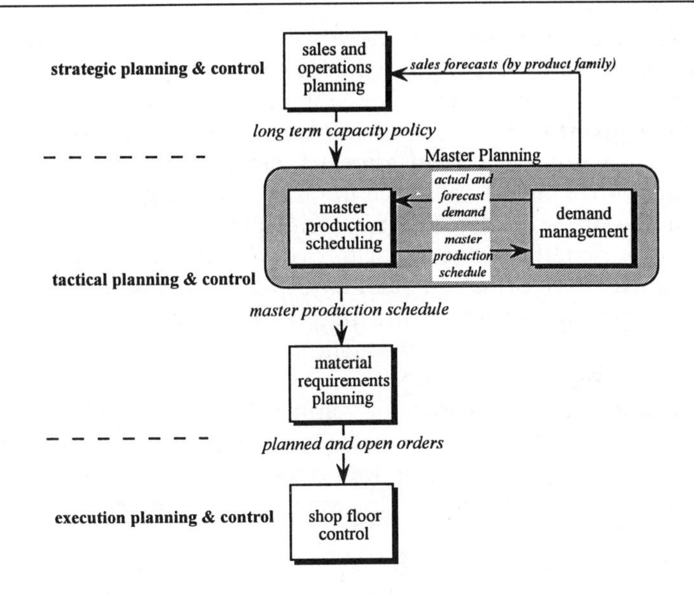

Figure 2.4 Different levels in MRP II-type systems.

Sales and operations planning is part of the business planning exercise and looks over a long-term (strategic) horizon. Its objective is to ensure that manufacturing capacity is adjusted to anticipated levels of sales. APICS defines the sales and operations planning activity as follows:

> the function of setting the overall level of manufacturing output (production plan) and other activities to best satisfy the current planned levels of sales (sales plan and/or forecasts), while meeting general business objectives . . . One of its primary purposes is to establish production rates that will achieve management's objective of maintaining, raising or lowering inventories or backlogs, while usually attempting to keep the workforce relatively stable . . .

The output of the activity is called the production plan and is defined as:

> the agreed-upon plan that comes from the sales and operations planning function, specifically the overall level of manufacturing output planned to be produced; usually stated as a monthly rate for each product family . . . The production plan is management's authorisation for the master scheduler to convert it into a more detailed plan, that is, the MPS.

The authors believe that the above definitions are misleading. The primary objective of sales and operations planning is not to define the production rates, nor is the scope of master production scheduling limited to converting the production plan into a more detailed plan, say by product within a certain product family. Sales and operations planning defines in the first place the manufacturing capacity that will or should be made available to meet anticipated levels of demand. The definition of manufacturing capacity over a long-term horizon may be accompanied by information such as expected manufacturing output, but this does not constitute the output of the sales and operations planning activity to master production scheduling. Information with respect to manufacturing output (may be expressed using a financial measure or a common measure of volume) is used to quantify the sales and operations planning process and to agree on targets between manufacturing and sales. The information may also be used as a yardstick against which to measure the performance of the manufacturing plant.

The real output of the sales and operations planning process towards master production scheduling should be viewed as a set of statements with respect to future enhancements or reductions of capacity. This set of statements could be called the long-term capacity policy. The long-term capacity policy defines the boundaries within which more detailed plans should be developed.

Master production scheduling is more than just converting predefined production rates to a more detailed plan. It is a critical planning activity that decides about the business response to market demand, within the boundaries of the long-term capacity policy.

The presence of the sales and operations planning activity is justified by the fact that most manufacturing companies can only adjust their levels of available capacity over a long-term horizon. Planning by quarter and at the level of product families is mostly satisfactory. The horizon may have a duration of up to 5 years.

2.2.3 TACTICAL PLANNING AND CONTROL

Tactical issues relate to the generation of detailed plans according to the guidelines and constraints issued from the business planning functions at the strategic planning level. In some cases, this may include a need to meet the demands imposed by a long-range production plan and the breakdown of the overall planned production in this plan into a feasible MPS. This MPS is then further broken down into their assemblies, subassemblies and components, and the creation of a time-phased plan of requirements, which is realistic in terms of capacity and material availability.

The tactical planning and control layer plans and controls purchasing and manufacturing activity for all (inventory) items in response to firm and anticipated demand. It is at this layer that the real MRP logic is applied.

Master production scheduling occupies a key position in the MRP architecture. It drives the MRP logic. The MPS expresses the manufacturing programme and is a set of plans for items assigned to the master production scheduler. Whatever the process of developing an MPS, the purpose of master production scheduling is to guarantee manual control over the manufacturing programme. This is in contrast with the plans that are developed automatically by the MRP process for all other items, not assigned to the master production scheduler. MRP can in this respect be considered as a pure and simple calculation engine. MPS is management's handle on the MRP process. It is only by means of master production scheduling that the response of the manufacturing organization to firm and anticipated demand is determined.

The horizon for tactical planning and control should be sufficient to enable development of plans for all of the items. Since planned order releases are offset from planned order receipts by the corresponding purchase or manufacturing lead times, and since orders are planned to satisfy requirements, which in turn derive by explosion of planned order receipts for the parent items, the minimum planning horizon should cover the cumulative lead time from release of purchase orders to realization of the finished products.

2.2.4 EXECUTION PLANNING AND CONTROL

Execution or operational planning and control issues essentially involve taking the output from the tactical planning phase, e.g. the planned orders from an MRP system, and managing the manufacturing system in quasi-real time to meet these requirements. The execution planning and control layer supplements the MRP logic with the required logic to plan and control the manufacturing environment on a day-by-day basis. The most important objective is to minimize lead times and work-in-process inventory, primarily by accurate control over which orders are released on the shop floor and how. The planning horizon for execution planning and control varies from a few shifts to a few days.

It is worthwhile repeating that the strategic, tactical and execution layers for planning and control each correspond with a certain time range and level of detail. Since planned order releases are offset from planned order receipts by the purchase or manufacturing lead time for the item, the planning horizon for an item is equal to the planning horizon for the parent item minus the purchase or manufacturing lead time for the item. As such, planning horizons shrink when descending the different levels of a bill of material structure.

2.2.5 THE INTERRELATIONSHIP BETWEEN PRODUCTION PLANNING AND CAPACITY PLANNING

Production planning as described above is very much interrelated with capacity planning. In Figure 2.5, different capacity planning functions according to the different production planning layers are presented. Planning horizons and level of detail for the capacity planning functions are the same as those for the corresponding production planning functions.

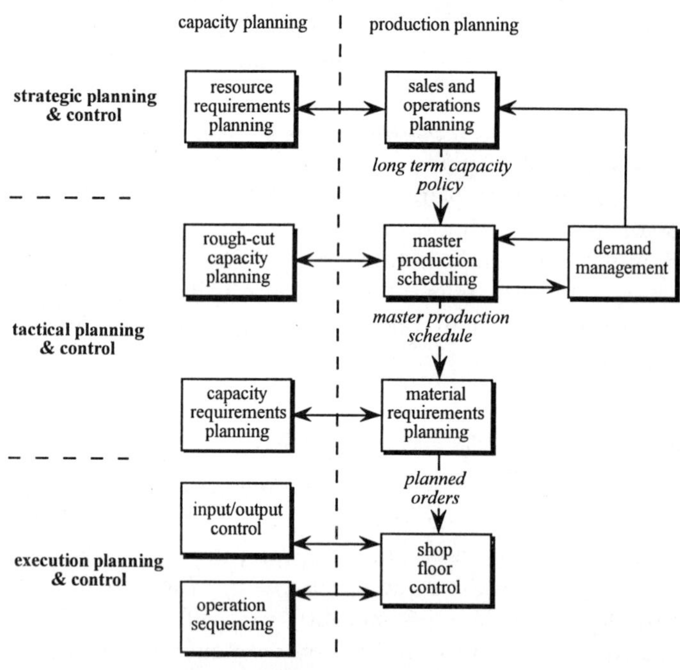

Figure 2.5 Capacity planning in MRP II-type systems.

The kind of capacity planning that is done for purposes of sales and operations planning is referred to as resource requirements planning. The importance of this level is to anticipate any required investments (land, buildings, machines, workforce) to be able to meet the anticipated levels of sales over a long horizon.

The capacity planning function corresponding with master production scheduling is typically called rough-cut capacity planning. The objective of this capacity planning function is to assess the MPS with respect to available capacity and the long-term capacity policy. The rough-cut capacity planning function is a critical one, since it constitutes the single test that a tentative MPS should pass before it can be authorized to drive the MRP run. Rough-cut capacity planning should not necessarily contain much detail (hence the descriptive term 'rough-cut'). However, it should be as reliable as possible for purposes of assessment of feasibility of the MPS.

Material requirements planning is typically accompanied by a capacity requirements planning function. Capacity requirements planning aims at calculating loads by work centre, based upon open and planned orders, as calculated by MRP. Capacity requirements planning is a pure calculation process and does not attempt to take account of the finite capacity of the work centres (capacity requirements planning is an **infinite** capacity calculation process). Its objectives are twofold. Firstly, it can be used to perform a check on feasibility that is more detailed and probably more accurate than the check by rough-cut capacity planning. Major discrepancies between required capacity and available capacity may mean revising the MPS. A second objective is to provide all information so that short- to medium-term measures can be taken to adjust capacity availability to capacity requirements. This is only possible if discrepancies are not too big and can be overcome, e.g. by using overtime or by scheduling shifts and holidays appropriately.

The traditional MRP architecture distinguishes between two capacity planning functions at the execution planning and control layer. The first one is input/output control, which serves two objectives. It firstly tracks input and output by work centre so that the necessary information is available to review the available capacity. Secondly, it tracks queue lengths and feeds them back to the shop floor control function, which can then attempt to control queue length by controlling the level of input. Queue control is necessary with MRP. Queues in front of work centres typically account for 95% of total lead time. If actual lead times have to stay within predetermined lead times, actual queue length has to stay within some predetermined estimates of acceptable queue length.

Operation sequencing is the second capacity planning function at the execution planning and control layer. Operation sequencing is a technique to schedule operations by work centre over a short horizon, while taking into account the finite capacities of work centres and according to a predefined dispatch priority algorithm. Input/output control, operations sequencing and shop floor control are currently integrated by the so-called Leitstand systems. This area is discussed in more detail in Chapter 5.

2.3 JUST-IN-TIME AND SYNCHRONOUS MANUFACTURING APPROACHES

2.3.1 JUST IN TIME

The just-in-time (JIT) approach has become very popular since the early 1980s. This is mainly due to the success of Japanese companies, in which the concepts originated in the early 1960s. The approach consists of three main areas (Browne *et al.* 1988):

1. JIT philosophical approach to manufacturing;
2. techniques for designing and planning the JIT manufacturing system;
3. techniques for control of the shop floor in a JIT system.

The philosophical approach to manufacturing proposes to achieve 'ideal' situations where companies have zero defects, zero set-up times, zero inventories, zero breakdowns, zero lead times and lot sizes of one. JIT can therefore be seen as an approach that aims at minimizing waste (inventory) and maximizing throughput, mainly by having materials supplied and produced 'just in time'.

JIT should be interpreted carefully. It can be interpreted literally at the supplier's side of the company: goods are called off daily and are being supplied just in time. However, JIT should not be understood in the same way at the customer's side of the company. Rather than supplying directly what is being ordered, most JIT environments rely upon intelligent management of the order book in order to achieve a satisfactory amount of stability and repetitiveness in the production programmes. The levelling and the forward visibility provide the necessary criteria to make JIT delivery possible. Deviations between forecast and call-off (mainly as regards mix of products) are absorbed by inventories at the supplier's side and by capacity flexibility within the company itself.

The purpose of master planning in a JIT environment is mainly to set a production rhythm (production rate) that optimally exploits the available capacity and meets projected customer demand. JIT manufacturing is about management of rates: the sales rate sets the production rate, which in turn sets the supply rate. As with MRP, JIT's focus is on flow of materials. The synchronization of flows is highly visible with JIT. As in the usual non-JIT environment, the MPS is also used to generate visibility not only for suppliers, but also for capacity planning internally. JIT could be described as being 'pull' driven by the downstream consumption and ultimately customer orders. It is not typically order driven as is typically the case with MRP II environments.

A **pull** system is typically associated with the concept of consumption-driven production. In the pull approach, the last operation is scheduled. In front of the last operation there are a number of standard containers of raw material (often called kanbans). The second last operation works as required to replenish the stocks in these containers. The second last operation in turn has a controlled number of standard containers of raw material in front of it, and the third last operation works as required to replenish these stocks. The work of all previous operations is controlled in the same way. The net effect of operating a pull control system is that manufacturing lead times are low and the amount of inventory in the processes controlled, thus avoiding quality and cost problems. Stocks of finished goods are best handled by a pull system, whereas intermediate stocks are undesirable in the short shelf-life factors involved in this industry, in which it is desirable to transform raw materials into finished goods as quickly as possible.

Push is the traditional and most common approach. It is mainly associated with order-driven production. It implies in its simplest form that the first operation in a process works to a schedule regardless of the capacity of the downstream operations. This can cause large queues between operations, and these in turn can give rise to long manufacturing lead times and quality problems (especially in the case of short shelf-life materials/product). These quality problems can cause additional inspection, rework and administration costs.

2.3.2 SYNCHRONOUS MANUFACTURING

This system has evolved in the last decade from being called optimized production technology (OPT), which basically involved a series of rules concerning bottlenecks and optimizations of machine utilization, to the concept of synchronous manufacturing. The main goal of the philosophy is to 'make money'. This goal is represented by three financial measurements: (i) net profit, (ii) return on investment and (iii) cash flow. These three financial measurements have three associated operational measurements: (i) throughput (rate of sales of products), (ii) inventory and (iii) operating expenses. The financial and operational measurements have three relationships which can be briefly described as follows.

1. An increase in throughput gives an increase in net profit.
2. A decrease in inventory gives an increase in return on investment.
3. A decrease in operating expenses gives an increase in cash flow.

In this approach shop floor issues, such as bottlenecks, set-ups and priorities, are treated in great detail. Ten rules are given, which, if followed, are claimed to give the expected increase in the making of money. The principles of synchronous manufacturing were the first step in the development of a production planning and control architecture that attaches more importance in the planning of capacity than materials. One of the goals of this book is to explain that MRP II is not in direct competition with the concepts of JIT or synchronous manufacturing, but the ideal planning framework should be designed and/or selected based on the requirements of the company in question.

In the following section an attempt is made to describe some of the drawbacks of the MRP II approach.

2.4 DRAWBACKS OF THE MRP II APPROACH

There are a number of problems associated with the MRP approach that make it unsuitable as a technique for manufacturing planning and control. The term MRP II was coined as a title for integrated systems approach for manufacturing resource planning, but this terminology kept the emphasis on MRP instead of the integrated system. The MRP approach is fundamentally flawed in, for example, the use of excessively long, predetermined lead times to backward schedule from established due dates, without taking into consideration the interaction of products competing for the same resources. In one extreme view, an MRP system could be simply classified as a pure information system without any real decision support capability.

As already mentioned, MRP II is just an extension of MRP features to support many other manufacturing functions beyond material planning, inventory and bill of material control. With this in mind, the assumptions that underlie the effective operation of MRP II should be noted (Plenert and Best 1986, Browne *et al.* 1988, Jones and Roberts 1990). There are many reasons advanced for the poor performance of some MRP systems in practice. Some of these relate to the need for widespread education in MRP thinking and to the necessity for top management commitment to ensure success.

Firstly, the problems that can be associated with any type of manufacturing planning and control system (MRP or non-MRP) are summarized in the following:

- Availability of basic and accurate data. It is assumed that the data will always be available and accurate. However, in many real-life situations companies do not use the data in the same formats and accuracy and maintenance of such a large data set can create many problems. This problem is more critical in MRP systems, as there is no priority given to data of higher importance.
- Complex manufacturing planning and control systems can be used. The complexity involved within many MRP systems is unnecessary for many smaller companies, and also the costs of implementation and consultancy/training can be overwhelming.
- Expertise needed to implement effective MPC systems exists. A basic assumption is that the expertise exists within companies to understand and work with advanced MPC systems and that training can be easily found.
- Disciplines required to implement MPC systems can be enforced. Sometimes, the disciplines and procedures necessary for the successful implementation of MPC systems can be very difficult for employees to come to terms with.
- The manufacturing environment and the procedures used can be modified to suit the MPC system. This is indeed a major obstacle in the implementation of many MRP systems. The manufacturing system is often adapted to suit the MRP system. The reverse may preferably be the case for many companies that have a good understanding of their manufacturing processes. The MRP system should be flexible enough to be 'customized' to suit the manufacturing environment in question if MRP is the proper approach.

Other drawbacks are more technical in nature and are specifically concerned with MRP systems. They are summarized in the following list.

- Lead times can be specified and (for optimal use) the overall product delivery time will be longer than the composite lead times of components.
- MRP uses inflated lead times in the explosion process. MRP indicates that orders should be executed sometime between the planned start and due dates. That is, the actual time intervals are shorter than the planned time intervals. Therefore the estimates used within a requirements planning system are inherently inaccurate. Each product is given a predefined production lead time.

These times are estimates, and unfortunately MRP users often treat these lead times as being very precise. In the current manufacturing environment, lead times are continuously decreasing and the customer interaction with companies requires that very accurate planning and lead time calculations need to be in place for successful customer satisfaction levels to be achieved.

- Design of the production environment, routeing and quality information. The areas of production environment design, the routeing information and attention to quality issues are not addressed in the MRP framework. MRP systems tend to assume that the environment exists as is and is not subject to change. This gives rise to the need for a production environment design element in the factory coordination layer of the manufacturing planning and control hierarchy. This will be discussed further in Chapter 5.
- Infinite capacity availability. MRP assumes that infinite capacity is available. For example, when a MPS is derived, all resources being used in the plant can be assumed to offer at least sufficient capacity to fulfil that schedule. This is based on the premise that the plan has already been passed through rough-cut capacity planning and therefore must be 'achievable'. Both JIT and OPT schedule production assume a limited capacity. In JIT the kanban card is used to control capacity and in OPT bottleneck scheduling is used. The MRP II approach has simplified the problems associated with the planning of capacity, and can be considered as a materials-oriented planning framework. However, this framework is still very suitable for many companies that have mainly a materials planning problem.
- Batch and lot sizing. Many implemented MRP systems tend to use the ideas of economic batch quantities after calculating the planned order quantities. Batches are larger than is necessary in order to offset the supposed costs of set-up and inventory. JIT and OPT have overcome the batch size problem. In JIT, the strategy is to reduce all set-up times to a minimum and OPT computes variable batch sizes.

These issues tend to show that the MRP core logic is a good fit for companies with a materials management problem, not for companies that depend heavily on the proper exploitation of capacity (bottleneck) resources.

Also, MRP attempts to address all the main time frames involved in manufacturing planning and control systems. It can work quite well at the tactical level for some companies, however it has overlooked the complexity of the strategic and execution planning levels. For example, in medium-term planning, average lead time values may be seen as adequate when developing master schedules and exploding into MRP planned order quantities. However, in the short term, finite scheduling techniques are needed to create finite schedules. In the long-term planning horizon, the use of planning bills of material is necessary for the calculation of forecasts and the generation of master production schedules, whereas in the medium term detailed bills of material are required in order to, for example, accept customer orders and perform pegging of critical resources and components. In general, more decision support systems are needed at the strategic level and extra scheduling systems are needed at the operational level. In the next section, a revised manufacturing planning and control architecture is presented, and this will be detailed further in the following chapters.

2.5 A REVISED MANUFACTURING PLANNING AND CONTROL ARCHITECTURE

In this section, the architecture presented in Figure 2.1 will be expanded upon. Detailed descriptions of the various levels are given in Chapters 3 to 5. At the strategic levels or business planning stage emphasis is placed on planning in terms of aggregated product families or products. The master production scheduling phase attempts to plan according to items that are shipped to the customer. Requirements planning is concerned with translating the MPS items into component requirements for short-term planning and purchasing. Factory coordination and production activity control mainly deal with the operations associated with manufacturing the component items and assembling the finished product.

2.5.1 BUSINESS PLANNING

Business planning deals with the long-term activities of a manufacturing organization. It develops the plans necessary to drive the sales, manufacturing and financial groups within an organization. These plans define markets to be addressed, products to be manufactured, required volumes and resources and the financial impact of meeting the overall objectives set by strategic planning within the organization.

Business planning will be discussed in detail in Chapter 3.

2.5.2 MASTER PLANNING

The master planning activity typically involves the development of an MPS, which is used to set the overall manufacturing programme for the company. Three key decisions must be made in the development of an MPS. These are decisions involving the following:

- What products to schedule?
- When to schedule them?
- How much of each product to schedule?

The first decision on **what** to schedule addresses the question of the unit to be used in the MPS, i.e. whether the MPS is to be based on end-level items, specific customer orders or some group of end items and product options. The answer depends largely on the manufacturing environment (MTS, ATO, MTO or ETO), in terms of the production approach used.

The second decision on **when** to schedule addresses the question of the planning horizon. The minimum planning horizon acceptable is that of the maximum cumulative lead time for the products. This lead time may include engineering design time, in the case of an MTO company, as well as material procurement time, production lead time and assembly lead time. (In some environments, the MPS may only cover subassembly lead times, with final assembly controlled via a final assembly schedule.) The planning horizon should extend beyond the cumulative lead time in order to provide visibility into the future. In deciding on the planning horizon, the aim should be to reduce the cumulative lead time in order to reduce the dependence on forecasting and to make the system more responsive to changes in customer demand. The decision involves the analysis of product lead times and customer delivery times.

The third decision on **how much** to schedule addresses the question of how demand for products is derived from various sources, and it may vary depending on the environment. In the manufacture of products to stock, future requirements are usually derived from past demand. In the manufacture of products to order, the backlog of customer orders may represent total production requirements. In custom assembly, a combination of forecasting and customer orders generates requirements. The organization of the distribution network and field inventory policy also directly affect production requirements. In many companies, demand comes from a number of different sources, e.g. customer orders, interplant transfers, inventory adjustments, service part requirements, backlog of actual orders, safety stock requirements and forecasts.

Master planning will be discussed in detail in Chapter 4.

2.5.3 EXECUTION PLANNING AND CONTROL

The MPS is one of the main inputs to the requirements planning process. The requirements planning system explodes the MPS using bill of material and inventory information to produce planned orders (Figure 2.2), without taking account of available capacity on the shop floor. The planned orders are converted into actual orders by the factory scheduling systems (factory coordination).

Production activity control systems attempt to control production within each cell on the shop floor, but have no capability at a plant level. Factory coordination can be used in an attempt to bridge this gap by using ideas drawn from, for example, JIT and OPT. The link with the design module within the factory coordination task provides the capability of structuring the production environment through group technology principles. Major benefits can arise from a regular and stable product flow that results from the use of such a production environment design system. The introduction of new products can be accomplished more efficiently and with less disruption to the process. The planning and control tasks concerning raw materials and work in progress become less complex. Owing to the stability of the production process, further production requirements for satisfying market demand can be determined with greater certainty.

Execution planning and control will be discussed in detail in Chapter 5.

In summary, we can state that in general there is a need for more detailed systems at the levels above and below requirements planning in the design of manufacturing planning and controls systems. Execution planning and control needs to perform its activities in a more detailed way and provide for the optimal use of resources at the shop floor level. In order to account for this added complexity there may be a need for both factory coordination and production activity control systems. Also, more intelligent systems are needed at the master planning level in terms of development of a feasible MPS. In the following section, an introduction to on-going extensions of the architecture to incorporate coverage of planning for the whole enterprise is discussed.

2.6 FROM PRODUCTION PLANNING TO ENTERPRISE PLANNING

In the manufacturing environment of today, there is much greater interaction between the customer and the manufacturer. This means that, in order to produce goods to customer specifications and provide fast deliveries, the manufacturer must be in close contact with both suppliers and customers. In order to achieve this improved delivery performance and decrease the lead times within the enterprise, manufacturers need to have planning and control systems that enable very good synchronization and planning in all the activities of the enterprise. In the past the response was usually to attempt to make the factory more efficient and responsive by developing computer-integrated manufacturing solutions. Nowadays, however, the challenge is greater and requires that a degree of integration takes place across the total value chain. The MPS function is one of the key functions in providing this link between suppliers, the manufacturer and the customer. The 'extended enterprise' concept is evolving in an attempt to model the relationships and the communications involved (Figure 2.6).

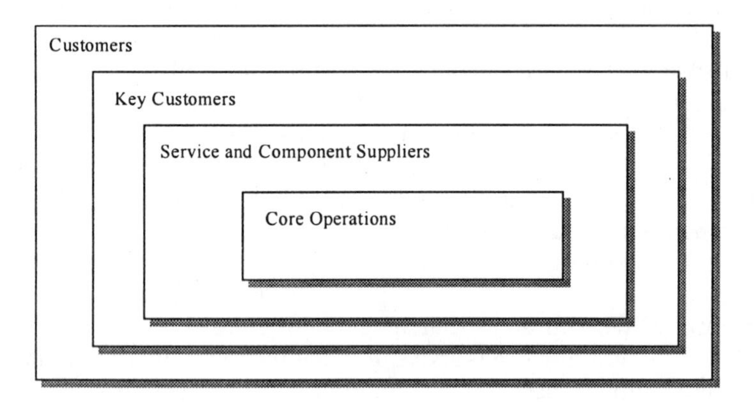

Figure 2.6 The extended enterprise.

ESPRIT project 6588, entitled Integrated Concurrent Enterprise Planning (ICEP), was developed to specifically investigate the expansion of the IMPACS project domain to that of complete enterprise planning. The ICEP consortium consists of Alcatel BELL, Antwerp (Belgium), Alcatel TITN Answare (Massy, France), Beyers Innovative Software, Antwerp (Belgium), CIM Research Group, CIMRU, UCG (Ireland) and SNECMA (France). The ICEP project team recognized the opportunity to enhance the MPS framework of IMPACS to include the activities of the whole manufacturing enterprise. This is the case in one of the project test sites, where the products, telephone exchange systems, involve engineering for customization and installation on the customer's site. Alcatel Bell and SNECMA are the ICEP project test sites, where the system is installed and tested. Each customer order is thus a project involving several activities in addition to manufacturing, and the various projects compete for the same resources. The ICEP project team developed ideas in relation to enterprise resource scheduling (ERS), in which pre- and post-manufacturing activities are scheduled together with manufacturing. An overall hierarchy representing the enterprise planning system is depicted in Figure 2.7. The topic of ERS is discussed in more detail in Chapter 4.

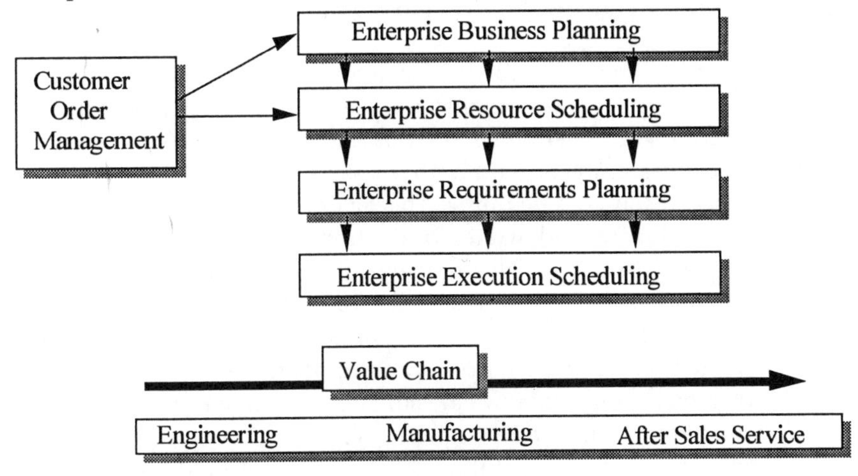

Figure 2.7 Integrated concurrent enterprise planning.

In Figure 2.7, the concept of the value chain is again visible. All pre- and post-manufacturing activities are adding value to the product to be sold. It is therefore very important that there is a proper synchronization between these activities. In effect, the whole system could be thought of as a 'customer to customer' cycle. The cycle begins with the initial contact from a customer and ends when the customer receives the finished product ready for use. This logically leads to the concept of concurrent enterprise planning, where ERS is a concurrent decision-making process managed by a planning board involving representatives of the engineering, manufacturing, installation/distribution but also marketing and sales functions. The planning board can be thought of as a team of individuals from different functional areas who each has some role to play in the development and verification of the enterprise plans. They are effectively the decision makers.

Overall, the functions of the integrated concurrent enterprise planning system are as follows:

- to plan the enterprise activities (engineering to order, purchasing, manufacturing and assembly, distribution, installation);
- to plan the enterprise resources (usage, availability);

- to enable a continuous improvement process of the enterprise system, through performance measurement and various analysis tools, based on cooperation between enterprise functions.

The different aspects of the layers depicted in Figure 2.7 are summarized in the following (mainly with respect to the project test sites):

- Enterprise business planning: This layer involves the forecasting of sales in terms of subassemblies and the scheduling of demand to take account of constrained final assembly capacity (expressed in terms of subassemblies per month).
- Enterprise resource scheduling. This layer consists of two sublayers, **master scheduling** and **master production scheduling**.
  1. Master scheduling. This involves the creation of a project plan per individual customer order or forecasted demand (this project plan includes all of the project activities, ranging from the tendering activity up to post-installation support); addition of project plans to arrive at estimated resource loads for each of the departments, from engineering, through production and on to training; and scheduling of project activities to take account of resource constraints within each of the departments.
  2. Master production scheduling. The MPS is only executed with production and is developed to satisfy the material requirements as can be derived from all firm and forecasted customer orders. The total requirements are the result of a totalling of the material requirements for each of the individual project plans. The MPS requirements for a certain project derive from the firm or forecasted configuration and are positioned by the scheduled production activity within the relevant project plan.
- Enterprise requirements planning. This is also only executed within the production cycle. The MPS requirements drive an MRP explosion in order to calculate planned orders for all internally produced and externally procured materials.
- Enterprise execution scheduling. This layer is concerned with the detailed shop floor scheduling logic needed within the production process.

It is clear from the above outline that the master scheduling function within the enterprise resource scheduling layer is the main activity in terms of integration across the entire value chain. Interested readers should refer to Higgins and Browne (1992), Dupas and Higgins (1993) and Dupas and Schepens (1993) for a more detailed discussion of the idea of concurrent planning and the ICEP project.

The concept of having one integrated system from business planning to execution planning and control is very challenging. The concurrent planning approach to incorporate the whole enterprise is even more challenging. The ICEP project has provided a very interesting set of solutions and an integrated enterprise architecture. However, the information flows between the different levels of the architecture need to be studied in more detail. More and more companies are beginning to look outside the walls of the enterprise and to develop planning and control systems that can deal with the whole supply or value chain.

There is a clear need for systems that integrate the planning systems of suppliers with the systems being used in the larger companies that require deliveries of high-quality materials in a JIT fashion. Suppliers need visibility of their customers' master schedule in order to predict their demand accurately and cope with short delivery lead times. The larger companies can also benefit from access to the master schedules of their suppliers so that they can, for example, electronically 'call off' the schedules.

In Figure 2.8, an architecture for manufacturing planning and control systems of large companies is depicted. This would be a typical architecture of a final assembler, which would be supplied by one or more suppliers. In this figure, there are two main external entities. These are the supplier(s) and the customer(s). Assembly takes place within PAC cells and is typically driven by both an MPS and a final assembly schedule. Orders may enter the system at different levels and are finalized in the final assembly cell.

In Figure 2.9, a much simpler architecture is depicted for the small manufacturing enterprises. [The manufacturing planning and control system for small manufacturing enterprises (SMEs) is discussed in Chapter 5.] Here the customer is the larger assembly company, which has a direct input to the overall production planning of the supplier. The aggregate production planning of the suppliers may well involve MRP, but as with any size of company the planning technique depends on the production environment and whether or not the company faces a materials or capacity management problem.

An attempt to merge these two figures (2.8 and 2.9) is shown in Figure 2.10. Here we see the supply chain, starting from suppliers and going right through to the final customers. There may be more than one supplier to a final assembly company and the links between the production planning and control systems could be achieved via the use of electronic data interchange (EDI) technology.

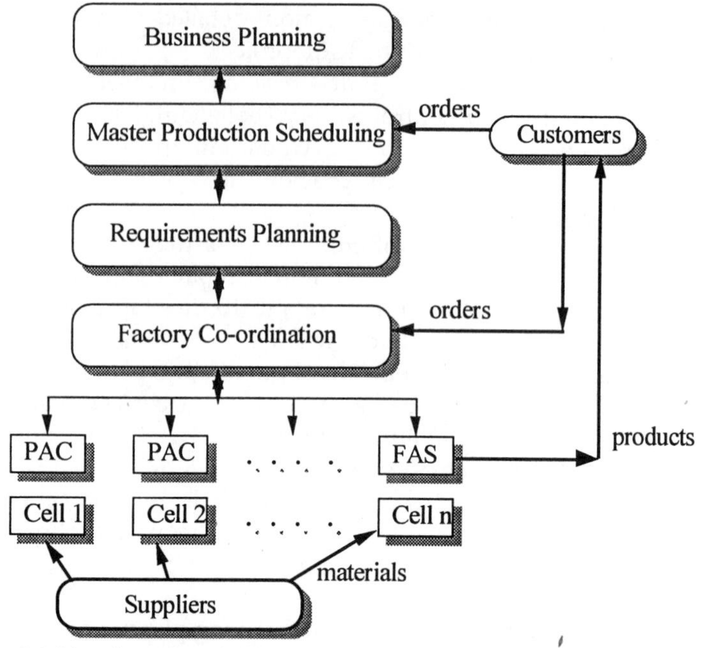

Figure 2.8 Manufacturing planning and control systems architecture of a large assembly type company showing customer and supplier (small manufacturing enterprise or contract company) interaction.

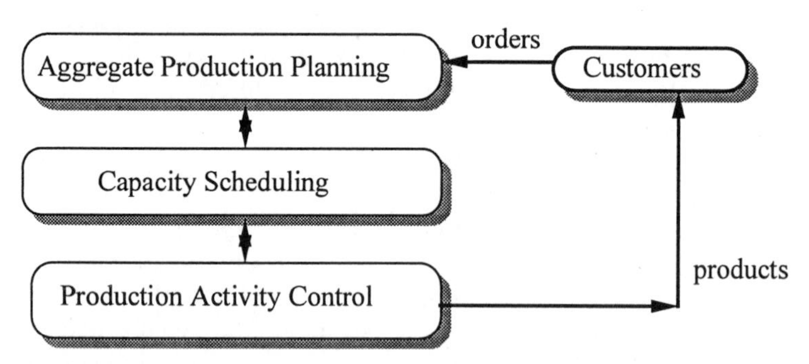

Figure 2.9 Manufacturing planning and control systems architecture for a small manufacturing enterprise showing customer (large assembly company) interaction.

This integration with suppliers and customers as part of the overall value chain or supply chain approach to planning and control can contain many overheads, for example the tremendous amount of data that must be managed and the integrity of these data. The Japanese have started to focus on the idea of lean manufacturing in their approach to manufacturing planning and control. Essentially, lean manufacturing involves using less of everything. They propose lean manufacturing as an approach that can help to reduce the time and effort to design the products, human effort and materials needed to make the products, defects found in the process and the inventories. There are five key principles involved.

- The maximum number of tasks and responsibilities are transferred to those actually adding value to the product.
- Smaller lot sizes, just-in-time and zero defects objectives combine to give an effective system for detecting, tracing and preventing defects.
- A comprehensive communications system means that everyone understands the workings of the factory and can respond quickly to problems that are encountered.
- Workers are divided into teams that are trained to do each job in their area.
- There is a strong sense of obligation between the company and its employees.

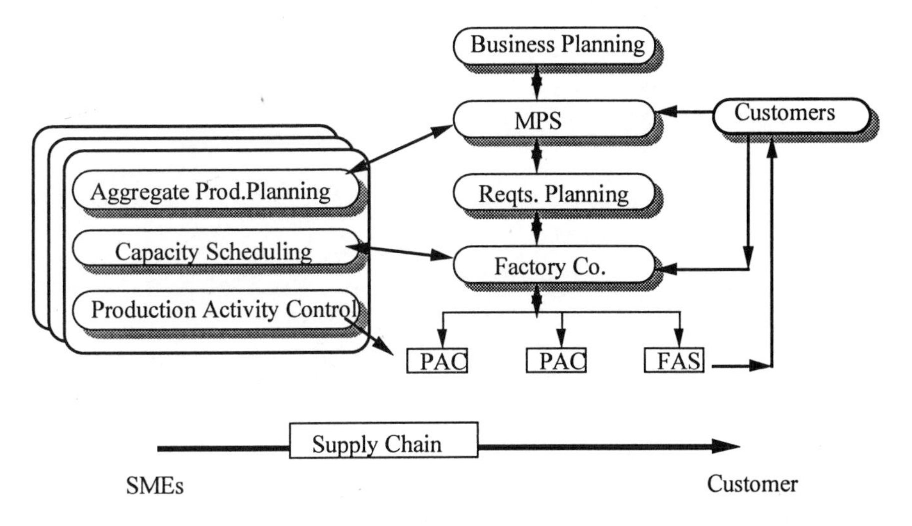

Figure 2.10 Combined picture showing interactions using EDI between suppliers, assembly plants and customers.

There are numerous factors proposed to differentiate lean from traditional manufacturing. Lean producers generally have one project leader responsible for the project from start to finish. The project leader organizes a small team from marketing, manufacturing and suppliers who are committed to the project for its complete life cycle. The team works on different steps in parallel rather than waiting for each step to be completed. This avoids time wasting. The small teams tend to have better communication and the suppliers benefit from extra information. In summary, lean manufacturing tends to be customer driven and involves the suppliers at very early stages in design.

2.7 SUMMARY

In this chapter, some of the main topics in the area of manufacturing planning and control systems were described. These descriptions included MRP II, JIT and synchronous manufacturing. Drawbacks to the MRP approach were also described and a revised framework for manufacturing planning and control as well as some concepts concerning enterprise and supply/value chain planning and control were briefly described. Detailed discussions will be given in Chapters 3 to 5 on business planning, master planning and execution planning and control issues.

Business planning 3

3.1 INTRODUCTION

The goal of a manufacturing business is to make profit. Consistent profit making can best be supported by management having a clear vision of how a sustained competitive advantage can be achieved and maintained for its business. Manufacturing business planning addresses the creation and implementation of this vision.

The task of creating and implementing the manufacturing business plan has traditionally been left outside the production and inventory management architecture. The master schedule or the shop floor control **tasks** can operate quite comfortably without the need for any **formal direct** inputs or guidelines from the business planning process. Nevertheless, decision making at the master schedule and shop floor control levels usually involves certain default assumptions such as:

- always accepting extra customer orders;
- using existing underutilized resources when introducing new products;
- accepting and scheduling all types of new customer orders;
- striving to increase the utilization of all factory resources.

While these assumptions seem sensible, they may not always be consistent with the business's longer term vision, resulting in an unnecessary waste of company resources or substandard service levels. Improvement philosophies such as just in time and total quality management (TQM) must be tested for fit with the company's strategic framework if they are to be successful in creating the type of competitive advantage envisaged in the business plan.

Manufacturing business planning comprises those decision choices concerning the planning horizon that extends beyond the product's cumulative lead time. Many businesses might argue that, because of this, its relevance to the day-to-day running of the manufacturing enterprise is minimal. In reality, the day-to-day constraints or capabilities of the manufacturing system are usually traceable to the strategic policies followed in the past. In a competitive environment, where responsiveness seems the only antidote to market pressures, an effective and integrated manufacturing business planning process is a prerequisite. The necessity of business process re-engineering (BPR) in the 1990s has challenged the traditional organization structures and behaviours, their supporting business processes and the performance measurements used. The structure of the business processes and of the supporting manufacturing infrastructure must be responsive to:

- the profit potential of the industry in which the firm is competing;
- the increasing expectations of customers;
- the emerging opportunities afforded by new technologies or human resource enablers.

Right-sizing cannot stop at an ideal size. Market opportunity is dynamic and temporal in nature. Organization structures are slowly evolving to accommodate this reality. An original motivation for the bureaucratic structure was to buffer the organization from the disturbances of the outside world. Competitive pressures and increased flexibility demands a move towards a flatter structure where communication with the external and internal environments is enhanced. The trend towards increased market pressure has continued. The continuity of lines of business (and indeed employment) is no longer an ideal to be maintained by the business. The focusing of the manufacturing enterprise, the breaking up of unprofitable lines and the increased use of benchmarking all characterize management's new respect for shareholder's funds. The implications are that aspects of the roles of fund management and brokering are infiltrating new management practice. Lines of business are no longer coupled to the enterprise with organizational glue but by the strength of their profitability impact. Below par performers are filtered out or targeted for re-engineering. Unhealthy product cross-subsidization is an immediate target but not the only one. To survive in this new competitive world, manufacturing management must have an accurate understanding of their

- resource needs
- costs
- service performance

at the product and activity level.

Two forms of management already exist. **Top-down management** is concerned with the business portfolios being currently funded. Product lines are continually evaluated against each other and the external environment. Many decisions must be taken, including the:

- choice of products (or services) to make or buy;
- level of capacity to be held;
- forms of capacity to be used.

The second form of management is **operations management**, which addresses the issues relating to the effective running of the production line. It is a bottom-up process. The performance criteria are prescribed by top management. Operations management is concerned with meeting these expectations while running the business in the most effective manner possible. The rate of performance improvement, the potential for improvement and the current performance are important ingredients in the glue that supports product line retention. The ability to track and predict resources, costs and service, as well as understanding the factors that influence them, is essential information for both forms of management. These tasks are not performed well in discrete manufacturing today. In this chapter, a **four-task model** is presented, which addresses the requirements for top-down and bottom-up planning. The approach to be outlined reflects the need for managers to understand and model their enterprises in detail, thus ensuring that informed manufacturing business planning decisions can be taken. The objective of this chapter is to outline some practical techniques and principles that can serve to guide and assist planners challenged with the task of formulating and implementing long-range plans within a manufacturing enterprise. Readers interested in the detailed content of a manufacturing strategy should consult the references included in the bibliography.

3.2 CREATING THE MANUFACTURING RESPONSE

Creating the manufacturing business plan begins with an understanding of the customer's requirements. The 'house of quality', the design tool of the management approach known as quality function deployment (QFD), provides a means for aligning manufacturing responses with customer requirements. This term emanates from the shape (i.e. a tiled roof) of the basic matrix used in QFD to correlate customer's requirements against product or service features (engineering characteristics). The 'roof' of the matrix is used to identify which of the engineering characteristics are in conflict with each other

(Eureka and Ryan 1988). Customer requirements or attributes are translated into targets for the key manufacturing processes, which in turn establish the production requirements. Customers' product and service requirements (attributes) can be bundled into an **order-winning criteria** classification. The criteria relates the market's varying preference for

- price competitiveness
- availability
- dependability
- quality
- features.

These preferences may change over time. It is also likely that a manufacturing business may find itself supplying a range of products or services to a customer base that has widely varying preferences. Management must proactively ensure that appropriate manufacturing capabilities are in place to support the product and service features demanded by customers (Figure 3.1).

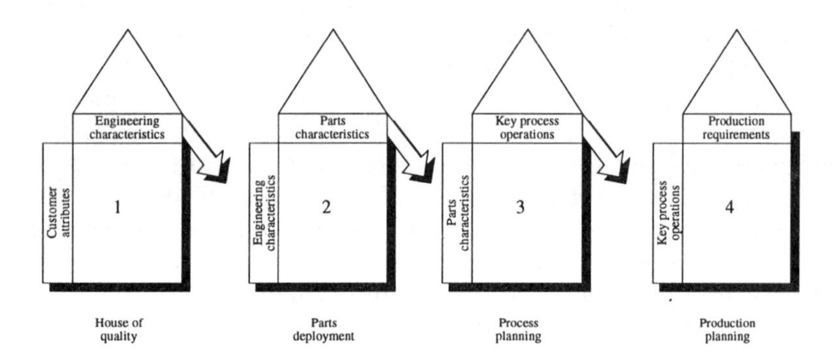

Figure 3.1 Linking the customer's voice through to manufacturing decision making.

It will not always be possible to respond and match all the dimensions of customer expectations. The customer may expect a highly dependable supply, but this may be difficult (initially expensive) to accommodate because of unstable manufacturing processes or complex supply logistics. Management may compromise by holding increased inventories of finished goods until its internal supply capability is sufficiently stabilized and improved.

As is evident, there seems to be a certain **degree of conflict** involved in responding appropriately to the different order-winning criteria of the marketplace. This apparent trade-off must be managed by the manufacturing system. **How it is managed is of little consequence to the customer.** Nevertheless, the customer's voice should influence the design choices and trade-offs that are made. Order-winning criteria are not independent of each other. Focusing on one critical success factor such as quality may have a negative impact on the other capabilities such as cost efficiency or flexibility. Capability development is a cumulative process. While short-term improvements can be achieved by focusing on one specific critical success factor to the exclusion of the others, it can be observed that these improvements tend to be short-lived. There is a certain preferred sequence involved in capability development. Some capabilities act as the building blocks for others.

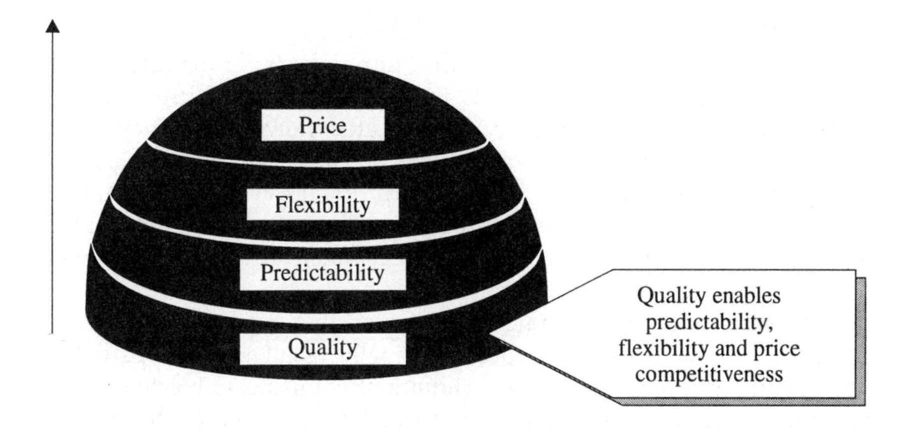

Figure 3.2 The sandcone model of capability development.

The 'sandcone model' (Figure 3.2) (De Meyer *et al.* 1989) proposes an ordered sequence of capability development. For example, dependability must be in place before responsiveness can be achieved. Quality processes and systems precede these and underpin all capabilities. There are many forms for each critical success factor. Becoming a flexible supplier could imply any of the following:

- developing the ability to accommodate unstable demand;
- being able to respond to dynamic supply constraints;
- always getting new products to the market early;
- responding to a rapid rate of product and/or process technology change.

Predictability in terms of consistent quality and on-time deliveries are the starting points for any manufacturing plan. Without these, business survival is in doubt. These capabilities can be regarded as the order qualifying criteria or entry requirements to the marketplace, but they do not guarantee success in the marketplace. The ability to outperform the competition and gain a competitive advantage must be developed. This is achieved through:

- price competition: offering a product that is cheaper than that of the competition

  or

- product differentiation: being more flexible than the competition or using new technologies more effectively.

The chosen strategy will have profound implications for manufacturing in terms of the choice of systems and processes to be used. A cost leadership strategy may encourage dedicated flows, low inventories, standardized skills and processes, a focus on capacity utilization and an emphasis on stabilized and formalized processes. The differentiation strategy may foster flexibility through a flexible and highly skilled workforce, informal systems, a focus on customer orders and the use of decoupling points and capacity buffers.

Without a well-defined strategy, decisions taken regarding the choice of production systems and processes usually result in incompatibility and redundant capacity. This non-alignment of business objectives with process and system capabilities is described as the stuck in the middle syndrome (Quinn *et al.* 1988). Developing new capabilities, in an ordered and integrated way, to match customer expectations can help to create a sustained competitive advantage for the business.

But the benefits can take time to realize. Management must decide on a strategic balance between:

- developing new capabilities

  and

- exploiting existing capabilities.

These two objectives serve to polarize the intent behind the strategic, tactical and operational layers of manufacturing planning. Shorter range planning is concerned primarily with the exploitation of existing capabilities. The primary focus of traditional master production scheduling and shop floor control has been towards the optimal use of existing capacities. Strategic planning addresses both objectives.

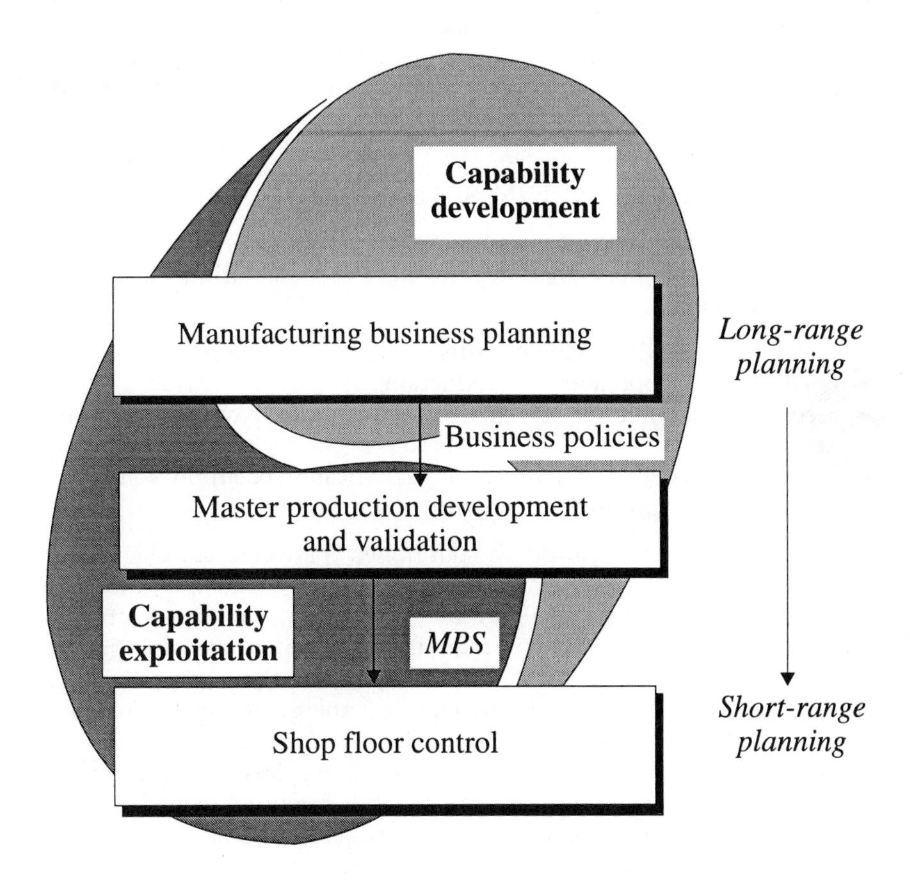

Figure 3.3 Capability development versus capability exploitation.

Manufacturing management must strike a balance between its development and exploitation plans (Figure 3.3). Capability development provides the future opportunities for tactical exploitation. For example, information technology could help to integrate company-wide business processes enabling delivery lead times to be shortened, quality of service to be improved and costs reduced. But the investments required for development programmes are likely to be funded from current capabilities and tactical exploitation plans. The just-in-time philosophy addresses both tactical exploitation and capability development. The pull process promises lower levels of inventory, but this is not practical without developing capabilities to reduce set-up times and improving synchronization. The instances where JIT implementations are seen as failures can be attributed to management dining à la carte from the menu of JIT prerequisites, concentrating exclusively on the exploitation or development aspects.

3.3 MANUFACTURING BUSINESS PLANNING

Five generic business processes or cycles have been identified which describe the work performed within most manufacturing businesses. The processes include:

- the booking of the customer order to billing the client;
- first to the last operation in production;
- request for quotation to the receipt of, and payment for, raw materials;
- specification of components to the identification of supply sources;
- design of new products to the full-volume production.

The manufacturing business is positioned between suppliers and customers on the supply chain. The performance and expectations of these groups influence the enormity of the task that manufacturing must address. The role that the manufacturing enterprise plays in the overall supply chain is constantly changing as the composition of suppliers, customers, competition and technology changes. This has internal implications resulting in the likelihood of change in the range of tasks performed within manufacturing. Some businesses may be compelled to, or prefer to, perform the full range of manufacturing activities. This form of manufacturing configuration tends to be more sluggish in its response to market changes. Other configurations may be less demanding, focusing on the repeatable tasks of order entry and production tasks, while outsourcing the remainder of its tasks. The configuration that is adopted will influence the:

- repeatability of tasks performed;
- overall response time of the business to external change;
- choice of manufacturing control system to be used.

As the range of tasks performed within the factory increases, the costs and complexity of managing the system also tends to increase. The design of the manufacturing system should be concerned with the **complexity** of the tasks performed as reflected by the amount of resource and level of skill required to perform each task. As the diversity of the activities performed within manufacturing increases, the frequency and repeatability of each tasks tends to reduce. This has a negative impact on organizational learning and the potential for improvement associated with larger scale production. The second factor that influences the design of the manufacturing system is the **uncertainty** of these tasks. Certain processes in the manufacturing system may be inherently unpredictable. These disturbances must be filtered in an acceptable way from the end customer. Finished goods inventory is often used as the buffer. The customer is paying the business to manage the associated uncertainty and complexity. The challenge for management is to reduce or eliminate the impact of these forces. An uncontrolled increase in the levels of complexity and uncertainty is likely to result in a deterioration in customer service levels as well as an increase in the level of resources needed to run the business. These two drivers of manufacturing costs and performance must be controlled through the manufacturing business planning process.

3.3.1 STEPS TOWARDS A MANUFACTURING BUSINESS PLAN

Order-winning criteria help to classify the customer's degree of preference for price, quality, dependability, flexibility and product features. A customer's perception of a supplier's performance along these dimensions may be real or perceived. Also, the value that different customers or the market places on these criteria may differ. The just-in-time philosophy encourages the pursuit of a customer selection strategy where 'good' customers are targeted. Good customers are those who value the efforts made by good suppliers. Market segmentation (i.e. targeting the potential customer groups) is an important input into the business planning process.

Having analysed and characterized the market requirements, manufacturing must decide upon an appropriate response. The response adopted will be influenced by the requirements of the market to be targeted, the requirements of other business functions (these depend upon the overall organization structure, but they usually include sales and marketing, corporate finance, research and development, etc.), the market share targeted, the current capabilities of manufacturing unit, the resources available, etc. The strategy chosen should reflect the response (types and levels of service to be offered) against each of the order-winning criteria (above) and also a ranking of these responses in order of importance. In effect, the basis on which the business will compete. This will serve to clarify manufacturing's focus of attention. Creating a vision of the manufacturing response capabilities is a good starting point in being a successful market player. **Manufacturing business planning** is the process of understanding and managing that vision.

There are two components to the manufacturing business plan – the **manufacturing strategy** and the **long-range production plans or operating/development plans**. The manufacturing strategy can only be formulated in conjunction with the other functional strategies such as sales and marketing strategy, technology strategy, etc. The objectives of a manufacturing strategy include defining manufacturing's role within the business and explaining how this role can be fulfilled. Manufacturing can be used as a strategic weapon for creating and sustaining long-term competitive advantage for the business. This can be manifested in a number of ways such as:

- the ability to bring a continual stream of new products to the market at an ever increasing rate;
- achieving a consistently high level of quality at every interface with the customer;
- providing dependable deliveries at all times.

The manufacturing strategy should indicate how the adopted manufacturing response to customer requirements can be achieved.

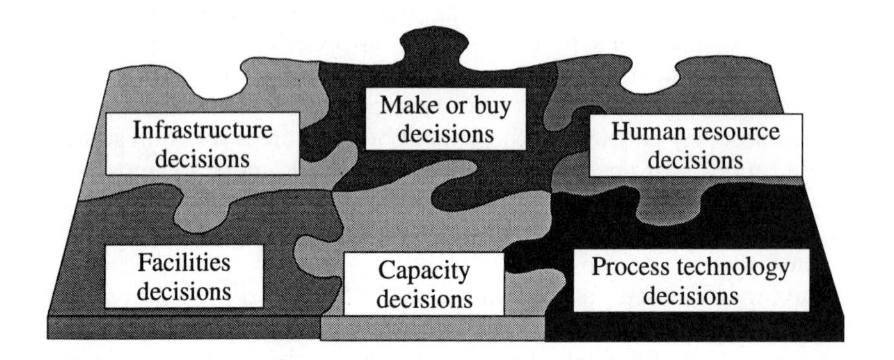

Figure 3.4 The six decision categories of a manufacturing strategy.

Maintaining or improving manufacturing performance is influenced by a variety of factors (Figure 3.4) such as the technology choices taken by the organization, the design of the organization, the choice of reward systems used, the complexity of the manufacturing processes, etc. The manufacturing strategy helps to formalize the selection of the set of business policies that address each factor in an appropriate and consistent manner. The strategy focuses on six decision categories. These involve the choice of:

- products or services to make or buy
- capacity policies
- facilities policies
- systems infrastructure and reward systems
- process technologies
- organization design.

The manufacturing strategy helps to develop the set of **guiding policies** that drives the manufacturing system. It consists of a sequence of decisions that, over time, enables a business unit to achieve a desired manufacturing structure, infrastructure and set of specific capabilities. The long-term production plan validates the set of guiding policies against current capabilities and the imposed business objectives. The output is a set of **proposed actions** outlining how the operating plans and continuous improvement activities are to be accomplished. The operating plans describe the resource requirements, targets and plans (output plan, purchase plan, headcount plan, space plan, etc.) of the business over the medium- and long-range horizons. These reflect the resource provisions that are anticipated in order to run the business in a manner consistent with the company's overall cost or differentiation strategies.

The guiding policies are communicated throughout the organization (Figure 3.5). Some are communicated directly to the MPS level. These include the:

- **capacity policy**, which establishes thresholds on the forms and amounts of capacity to be used;
- **responsiveness policy**, indicating the types and levels of response that are expected.

A formalized production plan is not an output from the manufacturing business planning process. But the necessary investments and guiding policies for the formulation of an MPS consistent with the established company objectives are. These investment guidelines and guiding policies serve as control limits for the master planning function in their formulation of the production plan. If the master planning function is unable to formulate a feasible production plan under these investment guidelines and policy constraints to realize the business objectives, then a re-examination of the manufacturing business plan is in order. Feedback between the master production scheduling and manufacturing business planning levels is achieved through a performance measurement system. This may take the form of a message stating that the business objectives are not being met or that the plan is no longer valid.

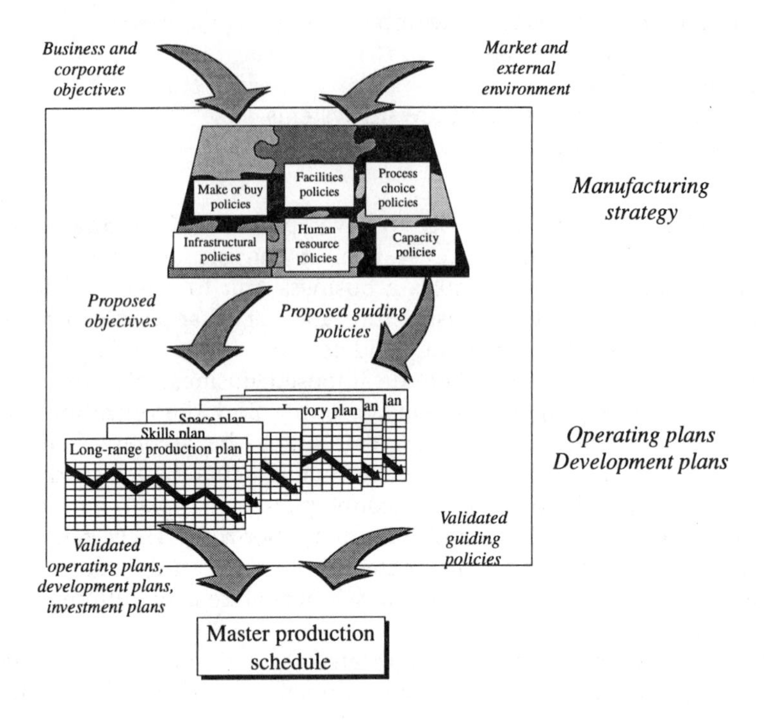

Figure 3.5 Investment plans and guiding policies.

3.3.2 EXISTING PRACTICE

While elements of continuous improvement programmes are part of most manufacturing business plans, their synergistic and integrative effects are rarely considered. Weaknesses exist in today's manufacturing business planning practice. Businesses assume that the introduction of improvement programmes such as EDI or JIT provides sufficient hedge against likely or indeed future unknown business changes. These programmes may, very well, be part of the overall solution, but they are no substitute for a properly formulated manufacturing vision and a plan for its implementation. The option of implementing these technologies and programmes is also available to competitors, enabling them to develop and replicate the same capabilities. It is unlikely that a sustained competitive advantage will be achieved by following this route.

The synergistic effects of a sequence of improvement programmes are rarely taken into account when formulating the business plan. The traditional trade-off model encourages investment opportunities to be evaluated against each other. Each candidate is regarded as being independent and in competition with each other. The candidate with the highest score is selected. The interdependencies of these investments should be understood and their cumulative impact modelled (Figure 3.6).

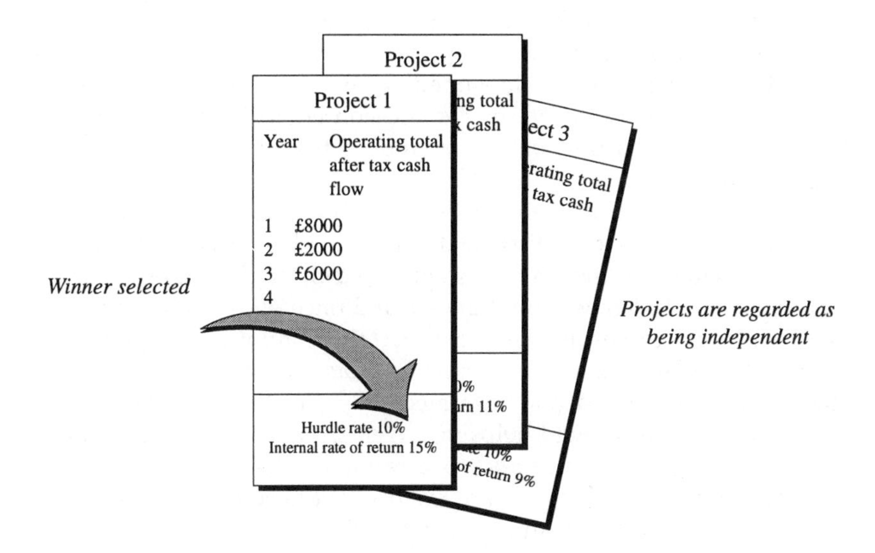

Figure 3.6 The trade-off model.

Programmes to improve manufacturing capability tend to be viewed mainly from a cost–benefit perspective. The traditional financial analysis approach prefers to focus on the certain returns likely to emerge from an investment. The uncertain returns and those that are unquantifiable receive a lower consideration. The conventional perspective is evident when an investment (i.e. the purchase of an extra machine) is considered (only) for its contribution to resolving an anticipated bottleneck or supporting a throughput plan. A more strategic perspective for such an investment would also factor into account the uncertain returns that might result, such as the opportunity to exploit the new knowledge that would result from the use of the new technology. The returns are uncertain, but the potential exists to exploit the learning to achieve new performance-enhancing capabilities, perhaps by making information more readily available to the sales force, thus, keeping the customer more informed. Process capabilities and not product features will determine the business's longer term competitive advantage.

3.3.3 THE NEED FOR A MODELLING APPROACH

A manufacturing facility is a very complex and dynamic system. It is insufficient to rely exclusively on standard financial criteria such as hurdle rates, etc., as the basis for decision making. Costs are influenced by the behaviours and capabilities of the manufacturing system. The planning process must be based upon an accurate understanding of how the specific manufacturing system works. Traditionally, key strategic decisions have depended heavily on financial analysis. Unfortunately, many factors in the decision choice have been ignored because of the difficulty in understanding their bottom-line financial implications.

Change in the form of BPR is encouraging organizations to look at radically new ways of satisfying their customers' needs, resulting in dramatic changes in the processes employed throughout their business. Although this change will be dramatic and once off, paradigms such as TQM can help to sustain and reinforce the continuous improvement efforts. The planning process must be able to drive and direct this change in a controlled and informed way. Feedback in the form of performance measurement helps management to focus attention on the key drivers of performance and, as a result, on the quality of previous decisions and on the areas of performance that need to be addressed for the future. Investments and guiding policies tend to have a cross-functional and synergistic impact across the organization. These likely impacts need to be factored into the planning process. An holistic view must be taken. Programmes should not be examined in isolation or on a departmental basis only.

It is accepted that the current capabilities of the manufacturing system are the consequence of the pattern of decisions taken in the past. It is insufficient to examine the implications of a single decision in isolation. The planning process should examine the cumulative impact of the possible sequence of policy decisions. Modelling should be an integral part of the business planning process. The challenge for managers is to build up their understanding of how their business really operates and to use this knowledge to make more informed decisions.

3.4 A FOUR-TASK MODEL

Every management team processes an agenda. This is composed of lists of changes to be implemented, investments to be made, opportunities to be sought, controls to be put in place, tasks to be accomplished, etc. We propose that management decisions should be addressed from four perspectives (Figure 3.7):

- resource planning
- cost control
- performance measurement
- policy making.

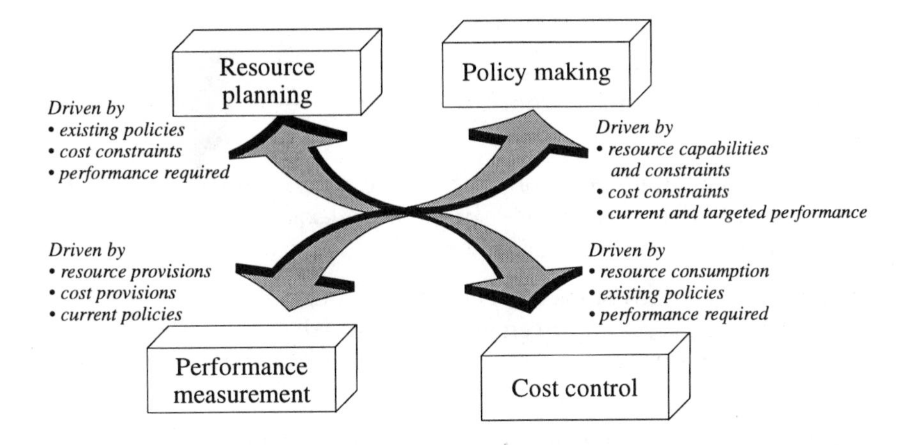

Figure 3.7 The four-task model.

These tasks are interdependent and should be considered together. Unfortunately, in practice, the management agenda is often based on poor information and tends to be incomplete. These perspectives tend to be considered in isolation and at different times. The subsequent sections address these tasks in more detail and outlines an approach that can be adopted which addresses the information needs of each perspective. Current practice is assessed and alternatives proposed. The four tasks should be viewed as being related aspects of all business planning decisions ensuring that the members of the management team are working from an informed and integrated view.

3.5 RESOURCE PLANNING

Being successful at the chosen form of market differentiation, in the longer term, demands an efficient use of all resources. Quality, predictability, flexibility and a low cost base as a supplier are achieved by having adequate resources in place at all times. Resource planning and control is one of the four identified tasks of manufacturing business planning. Some resources have long lead times, emphasizing the need for effective planning. A key task for management is to ensure that adequate resource provisions are made in a timely manner. Managers need visibility to future requirements in terms of the mix and the levels of resource required.

The distinction between managers' short-term decision needs and their longer term interests is also important. In the short term, the focus is on identifying and resolving resource bottleneck situations. The goal is to strike the correct balance between:

- avoiding bottleneck situations;
- maintaining an efficient operation;
- retaining flexibility.

However, the attainment of these objectives is influenced by the effectiveness of management's long-range resource planning efforts. Here, the planning effort is to determine:

- the resource provisions that should be made;
- what longer term improvements are required
- the likely resource and cost impact of alternative strategic policies.

3.5.1 THE NEED FOR A RESOURCE MODELLING SYSTEM

MRP II and its resource modelling capability could be used to calculate resource requirements, but its application tends to present oversimplistic results. MRP II was never really developed for use at the business planning level. It assumes that the levels of resource required to support the planned throughput varies directly as the volume of throughput changes. Bills of resource are created for the major product groupings, and these are multiplied by the proposed product volumes to yield the estimate of the level of the resource that is required. This assumption does not recognize the shift towards indirect activities within manufacturing. The level of activity in these areas is not directly linked to the volume of product shipped. It is also noteworthy that the emphasis is on determining the load position and not on the level of resource that will have to be provisioned. The fact that some resources can only be positioned in large increments is a factor that must be addressed off-line. For example, factory floor space may have to be added in modules of 100 000 square feet, which may provide space for between one and four extra productions lines. This resource is quite insensitive to small changes in throughput.

The traditional approach to resource planning assumes that the mix of load is unimportant. Synergy between different products are not factored. Each product or unit of output is treated as being independent of each other. Issues such as common set-up and similarity of processing are overlooked. The emphasis is on establishing a balance between supply and demand. The objective of maximizing throughput dominates. Objectives consistent with other order-winning criteria play a more passive role.

MRP II assumes that a uniform unit of measure is appropriate for expressing all the requirements for resources. Usually a measure such as standard labour hours or machine hours is used. The use of a standard measure of work is motivated by the wish to simplify planning. Using a universally standard measure assumes that all work is homogeneous and can be expressed in one dimension such as hours or material content.

The resource requirements within a function or department are driven by the practices and behaviours within that function as well by the practices and behaviours in interacting departments. For example, the ability of a final assembly department to meet its schedule is contingent on its own capabilities as well as the quality of service that it is receiving from feeder departments. Poor incoming quality and delivery dependability will serve to impair performance within the department. Modelling resource behaviour (Figure 3.8) should factor this cross-functional dependency of resource needs. The assumption that products or departments can be viewed as independent entities oversimplifies longer range resource planning.

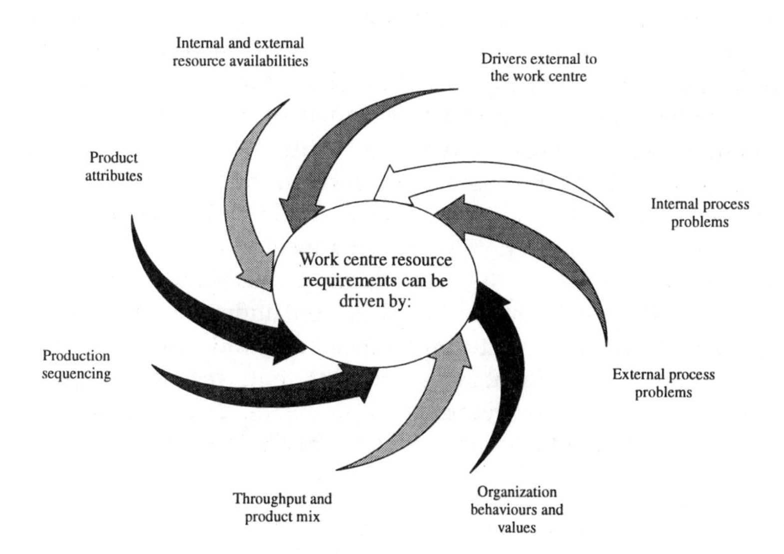

Figure 3.8 Some of the drivers of resource behaviour.

The goal of organization design is to effectively coordinate the tasks and activities of the various departments within the manufacturing system. This is characterized by two-way information flows between functions and by business processes that span many functions. To be effective, the manufacturing system must be well integrated. In contrast, manufacturing systems that are not working effectively are characterized by departmental problems and issues that manifest themselves across more than one function or department and between departments.

Paradoxically, this reliance on the integration of functions and departments is ignored in traditional approaches to modelling resource behaviour within manufacturing. Each department or work centre is assumed to behave independently. Multiplying a department standard (or process step standard) by the projected throughput results in an estimate of the load on that department, under the assumption that what occurs in other functions has no bearing on the department's requirement for resources. This is not a realistic assumption.

Resource increments or decrements should not be studied in isolation but instead in the context of their contribution to an incremental increase or decrease in overall throughput. This implies that the impact on the entire proposed collection of resources should be examined collectively and not individually, as is common practice. Recognition should be taken of the 'wandering bottleneck' effect. Addressing resource decisions from a localized perspective invariably leads to an overestimate of manufacturing's throughput capability, usually because the interdependence and synchronization of all the existing resource bottlenecks were not considered. This paints an overoptimistic financial perspective putting unnecessary pressure on the break-even point. The impact of proposed resource provisions should be studied in the context of the company's indicators of performance and its cost implications. These are all aspects of the same decision and should not be studied separately. Emphasis should be placed on understanding the current and projected manufacturing efficiency and in particular on characterizing the capabilities of different resource types. The performance of different resources in producing a varying mix and level of throughput should be understood. This is the compound effect of factors such as the:

- mix of resources used within the department;
- level of experience and training undergone by the personnel;
- effectiveness of the existing processes in use, etc.

These factors are complex and **ever changing**. The resource modelling system should help identify the drivers of performance and explain their likely impact on resource requirements. The model should also be concerned with identifying and quantifying capability improvement opportunities. Building models of how resource requirements behave with changing throughput affords managers the opportunity to understand how their business really works. Single-driver models based on standard labour hours or machine hours are inappropriate and can serve as a source of misleading information for management decision making. Resource modelling should take account of the policies governing the provision and use of resources. Account should be taken of whether the resources are fixed or variable. The level of resource used by the business may be greater or less than the projected resource requirement for the same time period. It can be observed that, even though the level of activity on the manufacturing line might decrease, neither the associated resources nor the cost may go away. For example, in the event that a small excess in factory space is forecasted, it is unlikely that management would dispose of this excess (for practical reasons). So, operating policies towards some types of resources govern the level of the resource that will actually be used. A fixed resource is a resource whose level is regarded as being fixed (constant) over time. The following properties exist.

- The level of the resource can remain constant if the requirement for the resource reduces. In effect, underutilization can be tolerated.
- The level of utilization of the resource can change as the activity level changes.
- Overload situations will require a review of the capacity policy.
- Most fixed resources are variable in the long run.

A resource is variable if its forecasted provision changes in response to changing throughput. Electrical power serves as an example.

- The level of the resource to be used will increase as the requirement for the resource increases.
- The level of the resource used reduces as the requirement for the resource decreases.

It is not always true that the need for resources is driven by the level of product produced. Often the use of a particular resource drives the need for another resource within that department or perhaps in another part of the manufacturing facility. Factory space and electrical requirements tend to be driven by the amount and nature of equipment used. The number of shift workers required might be primarily driven by shift policies and secondly by throughput requirements. These cause and effect relationships must be understood and captured in the modelling process. If resource behaviour can be understood and influenced, then the cost consequences can also be understood. This involves knowing the mix and levels of resources that are currently being deployed throughout the manufacturing facility and understanding what drives its likely behaviour.

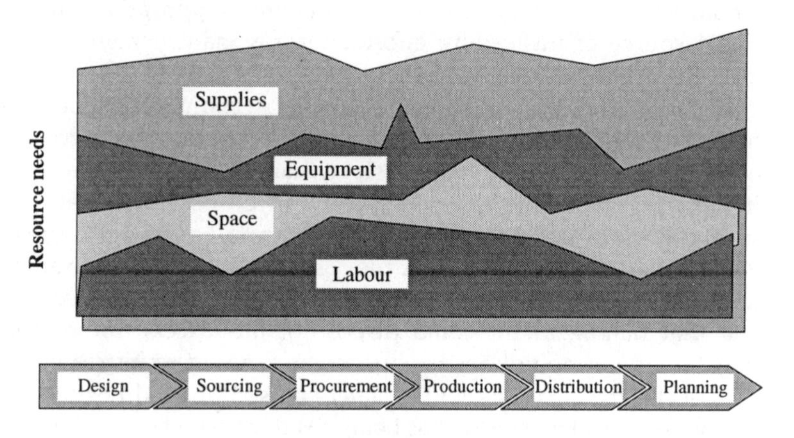

Figure 3.9 The resource landscape.

The model of the resource landscape (Figure 3.9) should scan all the direct and support activities that are performed. A global view gives visibility to all likely sources of constraints. Although this strategic view must be taken from an aggregate level, it does provide a sufficiently accurate barometer of the resource needs of the business. The theory of constraints approach applied at the master planning and shop floor control levels helps to ensure that these resource provisions are optimally utilized. The task of crafting the manufacturing response to the market requirements involves understanding the likely impact that proposed action plans and guiding policies will have on the resource landscape. For example, information concerning the likely impact of introducing a new process technology or redesigning the distribution channel should be understood in terms of the resource provisions that will be required. Planning should be seen as the ongoing process of gathering more information for improved understanding of the manufacturing system rather that the one-time task of deriving a static representation of the system. The continual pursuit of improvement programmes and the successful implementation of capability development plans should ensure that the resource landscape is being refined at an appropriate rate. This should be a focus point of attention for manufacturing business planning. Resource management and planning address the immediate as well as longer term aspects of resource use and requirements. Management's task is to ensure that adequate resources are positioned in a timely manner to ensure that the business's mission can be achieved, ensuring that the quality, predictability, flexibility and price expectations of the market are met.

3.5.2 RESOURCE PLANNING AT ST. COLUMCILLE'S HOSPITAL

Every hospital management team must strike the balance between the objectives of enhancing or maintaining the level of patient care with the ever increasing pressures to control costs. Although competition between hospitals is not formalized, inter-hospital comparisons are likely to become more important. The government's allocation of funding to health boards and direct to hospitals has become increasingly 'value for money' driven. The complexity of the hospital environment makes the management challenge more acute. Arguments such as 'teaching status' or 'complex mix of patients' are now accepted as being too imprecise for the purposes of explaining cost overruns or for attracting increased levels of funding. The Department of Health has responded by introducing the hospital inpatient enquiry system (HIPE) to most publicly funded hospitals in Ireland. It has provided a nationwide and objective basis for classifying the complexity of a hospital's casemix (the mix of hospital inpatients – equivalent to 'product mix' in a manufacturing context). Each hospital inpatient is diagnosed into one of 500 possible diagnostic related groups (DRGs). The HIPE system collects inpatient information, which is subsequently used to assign a DRG to each inpatient. The mix and volume of inpatients treated provide the basis for the government's funding allocations to hospitals. Funding standards have been established by DRG emphasizing the importance of casemix planning for the health boards and for hospital management teams.

In response to these pressures, the Eastern Health Board commissioned Digital to implement a system to help it to plan and manage the resources and costs in one of its hospitals. The resource planning system models the activities of the hospital, indicating the resource types and levels that are required to support the hospital's anticipated casemix. Cost implications are also built into the model. The management can monitor and model the complexity of its existing casemix as well as the resource efficiency of its operations. The system presents this information in a meaningful form to the hospital's management team, helping them to manage their forecasted casemix within their anticipated budget constraints. It has helped the hospital to harness hidden data that was previously inaccessible into valuable information for management decision making.

The organization design and decision-making structure found in a typical hospital environment tends to be different from those found in discrete manufacturing. Decisions and work undertaken by the hospital's medical and surgical consultants and their medical teams have a major bearing on a hospital's effectiveness and efficiency. Traditionally, consultants have not been provided with the necessary information to inform them of the implications of their individual casemixes on the hospital support services or on the hospital's overall cost performance. The resource planning system translates the hospital's casemix into activity levels and resource loads across all departments within the hospital. These resource implications can be monitored on a monthly or more frequent basis. Consultants understand their own casemixes. This serves to improve the data integrity of the HIPE system. The hospital's budget control process is supported by resource and cost information from the resource planning system providing timely indications of likely resource or cost problems. This enables the management team to take the necessary and timely corrective actions. A 'what if?' analysis capability exists to help identify the appropriate actions necessary to resolve likely resource or cost issues.

The model can be viewed as a simple demand/supply system. The first task is to identify 'what' the hospital's requirements are. HIPE casemix data serves as the input data for the construction of a forecasted bill of casemix. DRGs can be aggregated into 23 major diagnostic categories (MDCs). The bill of casemix (Figure 3.10) indicates the percentage frequency with which each DRG occurs within its major diagnostic category.

Periodic review of the hospital's actual HIPE reported casemix is used to refine and update the forecasted bill. The use of the bill of casemix helps the planner to differentiate a hospital's casemix from its inpatient throughput, which is calculated by multiplying the anticipated number of inpatients per MDC category by the bill of casemix for that MDC category. Forecasts are entered at the MDC level. In conjunction with this bill of casemix, a bill of activities is constructed for each DRG, to describe the range of direct and support services that constitute the treatment offered to the average inpatient within the grouping. Each DRG may differ in the range and level of hospital services involved and in the resources that are required to deliver these. The bill of activities is extended into a bill of resource by quantifying the resource types and levels required for each group.

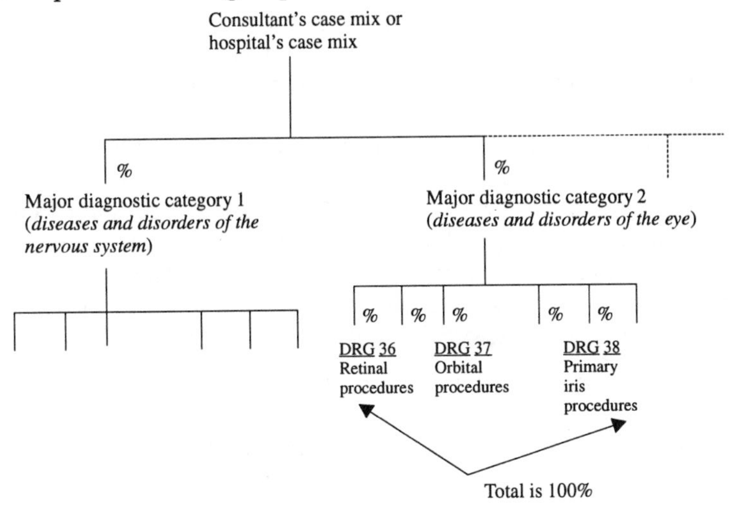

Figure 3.10 The forecasted bill of case mix.

The next step is to identify 'how' the hospital actually delivers these services (Figure 3.11).

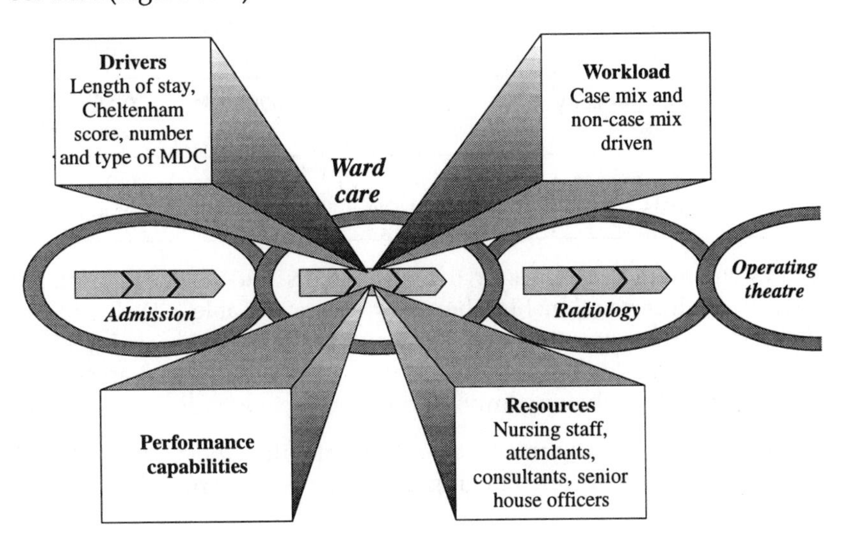

Figure 3.11 Modelling resources at the activity level.

The system focuses on modelling the business processes within the hospital, underpinning the notion that business processes drive a hospital's efficiency and effectiveness. It relates resource and cost figures to the day-to-day tasks performed within the hospital and to the way the hospital's internal processes are organized. Resource drivers and performance targets are identified by department. These drivers are periodically monitored throughout the planning year. Forecasted inpatient throughputs and/or 'what if?' figures can then be inputted to the model. The system reports the projected mix and level of resources that will be required to provide the necessary level of patient care to the hospital's anticipated inpatient casemix and throughput. The system is closed loop, capturing the actual cost for each major diagnostic category as well as the projected costs. It helps identify areas where cost overruns have occurred and the factors that have driven them.

Table 3.1 Estimating resources requirements

Department	Activities	Driver	MDC	MDC Workload	Resources
Laboratory	Biochemistry tests	No. of biochemistry tests	12	213 tests	0.04 biochemist
Laboratory	Biochemistry tests	No. of biochemistry tests	13	124 tests	0.02 biochemist
Laboratory	Biochemistry tests	No. of biochemistry tests	16	400 tests	0.12 biochemist
Laboratory	Biochemistry tests	No. of biochemistry tests	18	500 tests	0.15 biochemist
Laboratory	Biochemistry tests	No. of biochemistry tests	19	300 tests	0.08 biochemist

The system models the resource requirements and cost implications, showing the interrelationships between the two (Table 3.1). It reports the direct and support hospital services consumed by each inpatient category, the types and levels of resources involved and the associated costs. This has the following implications.

- The hospital management team can monitor their casemix and services performances, flagging possible cost overruns in a timely manner.

- The team has become increasingly customer and market focused, using the casemix information to drive the planning processes. Managers have been able to transform their budget process from being primarily expenditure focused to being a more dynamic, proactive process.
- Managers are in a stronger position to control their resource and cost budgets. Resource and cost trade-offs are identified. The cross-functional implications of changing workload and resource availabilities are understood and their coordination is improved.
- Operations can be planned more easily. Although its work is complex and its casemix dynamic, the future cost and resource needs of the hospital can, nevertheless, be anticipated. Changes in the type or in the number of patients to be treated are understood in terms of their impact on the hospital's workload, resources, costs and competitiveness.
- The management team is in a position to compare the hospital's cost-effectiveness against the Department of Health's standards. More importantly, managers have a deeper understanding of their hospital's internal processes, its cost implications and the factors and behaviours they can address that will have a bottom-line impact on their cost and resource performance.
- The necessary information to ensure that improvement efforts are properly directed is available. The cost and resource models inform managers of the bottom-line cost impact of their proposed changes (Figure 3.12).

By analysing its business processes and the average costs for each of its major diagnostic categories, the Eastern Health Board is in a position to compare the effectiveness of its various hospitals in managing some or all of its casemix for the region. Inter-hospital comparisons are possible. The Health Board and its hospitals can use the resource planning system to compare the likely cost implications of its anticipated casemix against provisioned funding levels both from the Department of Health and from its private patients. The budget holders are in an informed position to agree on the appropriate **level of activity** and the **cost provisions**. Likely shortfalls can be identified and corrective plans put in place in a timely manner. The use of the system has enhanced patient-focused management practices, helping the management team to coordinate its resource and service mix to treat the ever-changing mix of patients within their care.

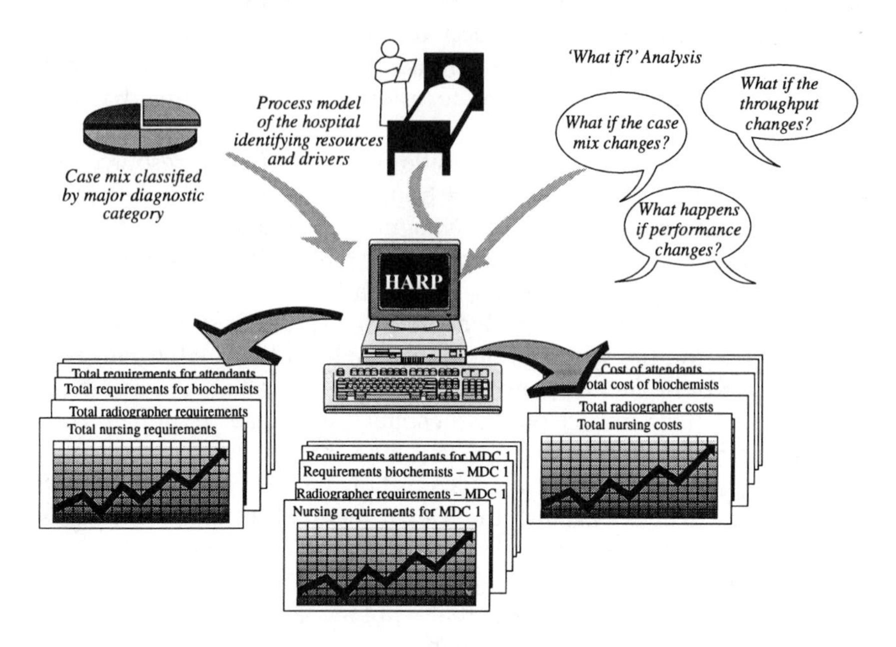

Figure 3.12 Resource and cost management in a hospital environment.

3.6 COST CONTROL

The second task identified in the four-task model is that of cost control. Most manufacturing organizations retain strong elements of a bureaucratic structure. Mintzberg (1983) described the typical bureaucratic organization as comprising of five elements. In this structure top management rely on the 'analyst' group, comprising engineers, accountants, personnel, etc., to regulate and control its business processes. This is achieved through the division of labour, design of processes and procedures, the imposition of control systems such as costing and quality systems, etc. This has resulted in a business composed of disparate professional influences. Although these different perspectives (such as accounting, quality, industrial engineering etc.) strive to achieve the best outcome for the manufacturing business, their proposed approaches also result in cross-functional inconsistencies. The management team has relied on these groups to implement the desired manufacturing system and on the feedback from these groups to understand how well the business is performing. Understanding the intricacies of manufacturing has tended to be a delegated task with an over-reliance on the traditional reporting system. Delegating the design tasks within manufacturing to 'analyst' groups has resulted in a great similarity in the way manufacturing companies across different businesses are measured and controlled and also in the problems that they experience.

Cost control has always engaged management attention as the primary tool for control of the business. It has provided the key information upon which most management decisions are based. Although it is intended to answer all questions about the business in an objective way, many more questions are raised. Traditional thinking in relation to cost control endeavours to analyse past and forecast future cost behaviour, but costs do not always behave according to these expectations. The observations below challenge the assumptions of the traditional view.

- Costs do not always go away when the level of work reduces.
- Costs do not always reduce when outputs fall.
- Costs can remain uncontrollable even when a strict variances management process is in place.

The nature of the market and the responding manufacturing system tend to be in constant change. New products are being introduced, new services being offered, new technologies applied and new customers found. The manufacturing system changes in response to these pressures. Investments are made in new process technologies, new skills are acquired and practised, new facilities added, etc. Management must steer the business successfully through all this change and manage the associated risk and cost implications ensuring that its cost structure is aligned with the market and stakeholder's expectations. Controlling and sometimes cutting costs is an ongoing task. A uniform percentage cut in costs across all functions (or departments) within the business usually reflects a lack of information on management's part. Sometimes, the target is the largest cost category that can be identified. The criticality of the identified function, its impact on other functions and the likely impact of the proposed cut on the indicators of performance are not always understood.

3.6.1 VISIBILITY TO COSTS

The management team must understand the impact of complexity and uncertainty on their business if it is to be in a position to make informed decisions. There is a natural tendency in business to evolve towards increased complexity. Perhaps this is primarily driven by the market's requirement for increased levels of customization, which, if responded to in a careless manner, can lead to an increased diversity of products and activities. Businesses have responded to their current plight by recognizing the need to become more focused in the activities they perform and services they provide. The task of understanding cost behaviour is made more difficult because of the level of uncertainty that exists within most businesses. Market requirements and supply capabilities can change and planned and unplanned events occur. These must all be factored into the assumptions and models. Without accurate and timely information, manufacturing can find itself operating from a financially disadvantageous position. A good model should indicate to management if:

- the wealth of the enterprise is being eroded by focusing on the wrong products and services;
- prices are being set below affordable levels;
- the running costs for the business are excessive for the future planning periods.

The answer to these questions lies in the understanding of:

- the costs to deliver a product or service;
- the behaviour of costs and how they are influenced by various identified drivers.

Good information is a vital ingredient for management's decision making and control tasks. Traditional approaches to costing are not sufficiently sensitive to the changed nature of the manufacturing business.

3.6.2 THE TRADITIONAL APPROACHES OF FULL AND DIRECT COSTING

An understanding of what products and services are costing to deliver is a prerequisite for effective decision making in business. There is an unfortunate tendency to accept increasing levels of demand and work as being 'naturally' good. Indeed the actual response to a request for increased output is usually positive. The traditional view in business is that increased levels of work should lead to increased utilization of all resources, which in turn should have a positive influence on unit costs and hence on profits. But, changes in the level of throughput are not necessarily mirrored by changes in the cost of running the business. Costing systems have been used as the primary instrument for management control in most businesses. **Standard costing,** the variation most widely used, has been around since the 1920s. Remarkably, no major advancements to the approach have taken place since its introduction despite the fact that major changes have occurred in the way most businesses operate. The focus of the approach is on the determination of product costs.

Managers have used feedback in relation to the flow of costs in their business to provide guidance for decision making and control. Decisions to be taken in relation to pricing and throughput volumes require managers to understand their product costs. Two approaches have been pursued. **Direct costing** identifies those cost elements that can be directly linked to the products or services being outputted. These include direct labour, direct material and direct expense. These are referred to as prime costs. But a significant portion of costs are not directly attributable to products and are not analysed and understood in the required detail. **Full costing** attempts to attribute all costs to all the products and services outputted. So each full or absorption cost of a product or service will be composed of both fixed and variable costs. The indirect costs of running the business are forecasted assuming specific throughput targets. This pool of cost is regarded as being required to support the budgeted product or service volumes, each of which is asked to absorb its 'fair and realistic' portion. If output falls short during this period, costs are said to be underabsorbed. The emphasis is on meeting the budgeted output even though this is often the wrong course of action, leading to excess inventories of the wrong product and resulting in wasted resources and capacity.

The implication of using full costing for decision making is that incorrect conclusions can easily be drawn. It assumes that the task of producing or supplying a product or service is composed of a series of homogeneous activities of similar complexity and cost behaviour. Traditionally, it has assumed that the amount of direct labour (or material) involved provides an accurate indicator of the cost of supplying a unit of the product or service. This assumption was used as the basis for allocating the indirect or support costs. But products that are regarded as being similar based on this criteria can often require dramatically different support activities. The consequence is that standard costing underestimates the cost of supplying a complex low-volume product while overestimating the costs of supplying a standardized high-volume product. A costing system that uses labour (or material) as the single basis for allocating support costs is termed as a **single-driver system**, i.e. product costs are assumed to be completely driven by the behaviour of direct labour or direct materials.

Compounding this reporting bias is the realization that these elements (more correctly direct labour) are reducing in relative size in most manufacturing businesses owing to the increased levels of automation and the associated growth in support activities required. This evolution makes the pro-rata approach of full costing less accurate. The implication for policy making in business is that certain courses of action are being pursued under false beliefs. The use of product costs derived in this manner can encourage management to set prices at an unsustainable level, to offer the wrong products and services and to retain the wrong processes and technologies (Figure 3.13).

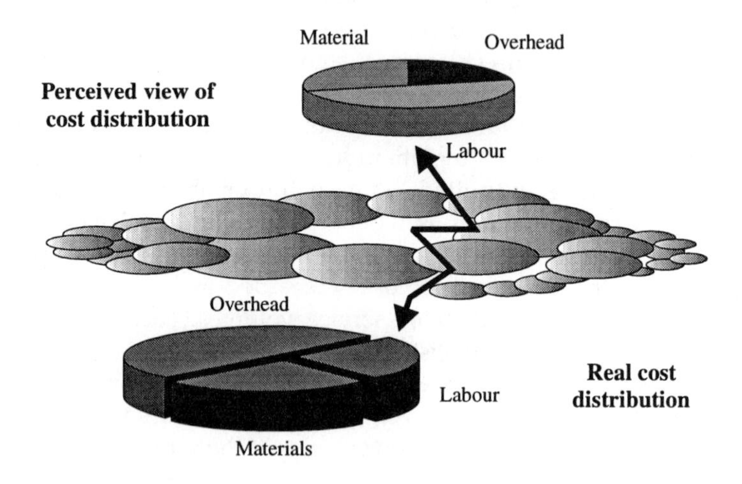

Figure 3.13 The information bias presented by activity-based costing.

3.6.3 CURRENT THINKING: THROUGHPUT VERSUS ACTIVITY-BASED COSTING

Activity-based costing (ABC) is a more recent approach to cost control. It has overcome the weaknesses of single driver cost systems and as a result serves to eliminate some of the reporting bias. Under ABC, the use of multiple drivers is possible, reflecting the fact that costs are driven by many factors i.e. the volume of the product, the product design and the quality levels being achieved, etc. This is more representative of the complex distribution of costs that occur in a real manufacturing environment. The analysis is top down in approach. The process begins by determining the expenditure for the last budget period. This is broken down by function or department and ultimately to activity level using an allocation process. The allocation percentages are determined by an interview process, resulting in an acceptable estimate of the costs of performing each activity during that period (Figure 3.14).

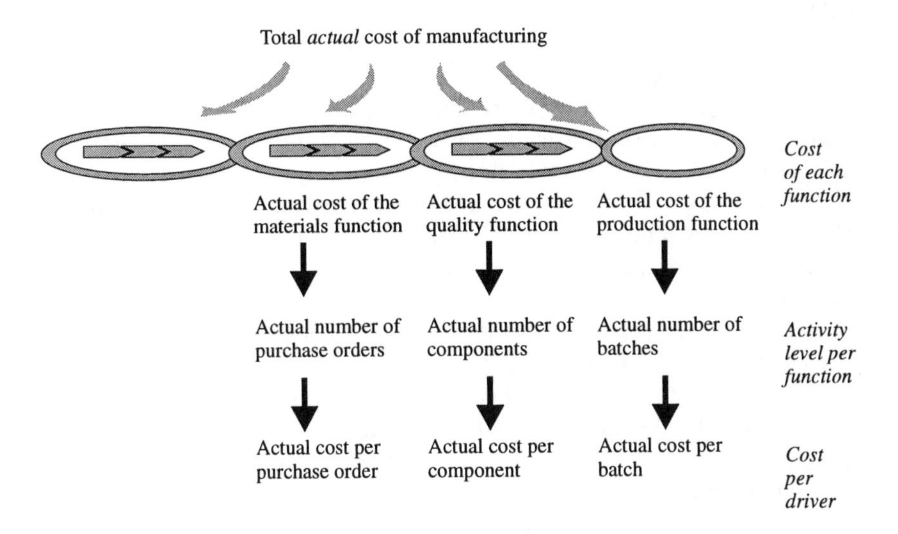

Figure 3.14 Simple ABC process.

The next step is to identify the cost driver for each activity. For example, the number of purchase orders may be identified as the driver of costs within the materials department. When the total driver value is calculated (in this example, the number of purchase orders placed last year), activity-based costing determines the actual cost per driver. The forecast for each driver (for the budget period) is multiplied by the actual cost for that driver to predict future costs associated with the driver. The sum of all forecasted costs will indicate the likely total cost of running the manufacturing system for the planning period. The underlying assumption is that costs change as the driver volumes change. Activity-based costing is a positive advance on the traditional approach to costing. It recognizes that the cost behaviour of the manufacturing system is complex and has many influencing factors that can be modelled by using a **multiple driver** approach. It overcomes the assumption that all costs are product volume related but can also be driven by non-volume factors such as batch size, product quality, level of product standardization, etc. The ABC approach provides a more accurate estimate of product costs and hence provides better quality information for decision making. It is more proactive than the traditional approach to costing. It tries to identify factors that drive and influence costs. Although it represents a major advance in thinking, it does have some weaknesses.

ABC assumes that the costs incurred will change as the level of work changes. Costs are predicted by multiplying the driver forecast by its derived cost for that driver. In most cases, the driver tends to be a measure of the level of work being performed. The underlying assumption is that as the levels of activity (drivers) change, costs will change accordingly. This does not take into account the current mix of resources at each activity centre and their existing levels of utilization. For example, a labour-limited activity centre may be working at full capacity and perhaps showing a low cost per driver unit, in comparison with a highly automated environment performing the same activity that is working at an underutilized level. Costs should change at a slower rate in the latter scenario. Activity-based costing does not take into account the likely constraints at each activity centre. The size of capacity increments will determine the rate at which costs will increase. Machine capacity is increased in terms of full machines. It is unusual to add 0.56 of a new machine. The rate of cost growth or reduction is not necessarily linear or directly related to changes in product volume or activity levels.

3.6.4 RECOMMENDATIONS

Costing systems tend to be used in two modes. The Western approach is focused on finding a satisfactory way of explaining the costs incurred in producing a product. The system is based on the principle of **feedback**. First, production takes place; afterwards the costs are analysed. The costing system tries to explain the past, but tends to do so on the basis of the direct labour or material involved in the production. This is a risky assumption to make. On finding an explanation for the past, it then assumes that this formula can be used for predicting the future behaviour of costs. The involvement of production line personnel is constrained to passive roles such as the logging of work order information. The analysis tends to be done in the language of cost only. Although costs are a consequence of resource consumption, the costing system does not concern itself with the factors that drove the requirements for resources. The focus is on the symptom (the cost elements) and not the underlying cause (the factors that drove the need for the resource). Of equal concern is the risk of product cost data being misapplied to predict future cost behaviour. For example, the likely marginal cost of outputting extra product is not obvious from the use of a product cost standard.

The other and more proactive approach to cost control is based on the principle of **feedforward**. This system attempts to establish cost targets that must be achieved if the business is to realize the objectives supported by its manufacturing strategy. The focus is less on precision. Instead the cost control system is used as a vehicle to drive business improvements. The objective is to challenge the system. The choice of the methods used to allocate overheads serves to direct behaviour within manufacturing. The choice of **the number of non-standard components** as driver method of allocating costs can serve to reduce sourcing as well as assembly costs. The use of material costs as the basis for allocation can serve to drive procurement costs down. The past is not assumed as an excuse for the future. The future is regarded as plannable. Cost targets are established that challenge the system to improve on previous performances. Management must ensure that the planned bias in the costing system in use does not outweigh the need for accurate information for decision making. If the decision is taken to allow a deliberate bias in the reporting convention used, then management must be sure that the intended manufacturing response is the appropriate one for the manufacturing business. The use of standard labour hours as a means for allocating overhead is designed to reduce the labour content of manufacturing. This may not be the appropriate direction for the business to take if flexibility and features are the requirements of the marketplace.

3.6.5 A RECOMMENDED WAY FORWARD BY MODELLING RESOURCES AS AN APPROACH TO COST CONTROL

It is important to realize the difference between having accurate estimates of product cost and having precise estimates. Traditionally, managers have devoted a significant effort to measuring their product/service costs. Variance against standard (estimates) has been the primary instrument of management control. The variance management process consumes a significant amount of management time and effort. Unfortunately, the standards used tend to be inaccurate and insufficiently dynamic. Significant trends are not reflected in a timely manner. Cost problems that are a symptom of a more underlying process problem are usually manifested long after the initial problem occurs. Nevertheless, businesses spend much time and effort determining and measuring conformance to these standards. A more practical approach is to get the balance between accuracy and precision right. It is not essential to know all possible variances against expectation, but it is essential to understand cost behaviour to be able to forecast costs with respect to anticipated changes. An understanding of resource requirements and likely resource behaviour will yield a valuable insight into cost behaviour. This does not require meticulous precision. The likely behaviour of resources and how this reflects itself into cost consequences should be captured. The main premise is that costs are driven by the level of resources consumed. In order to drive costs downwards we must understand and control the factors that influence resource consumption.

Business competitiveness is influenced by the manner in which the business deploys and uses its resources. This must be understood and modelled at an activity level. Companies having similar levels of output may incur completely different levels of cost. The reasons for this may be that the two business may be configured differently, they may manufacture the same products differently or the products manufactured may be slightly different. Management should progress away from a reliance on analysing its business through the data produced from traditional reporting systems as designed by 'analysts'. The practice of evaluating the business through ratios and variances only tests the manufacturing system for conformance to expectation. Instead, managers should use the opportunity to learn as much as possible about how 'their' business operates and how they can improve it.

Modelling is an integral part of the management task. All managers employ models as their aid to decision making, whether they are explicit or implicit models. Some managers practise 'management by walkabout' to improve their understanding of what is really going on in their factories. Managing a complex and dynamic business is impossible and impractical without such supports. Modelling the business is a required management task. It is important to understand how the business will react to likely changes in its environment, such as the introduction of new technologies, new processes etc. Modelling is a means of gaining a (collective) common understanding of how the business really functions and the various factors that drive this behaviour. It is even more important if the approach to model building is bottom up, where the model is developed from the inputs of the various functions within the organization, thus ensuring that there is an increased likelihood of a better understanding of the predicted implications and an acceptance of the adopted policies that are formulated in response to these. Models can forewarn the user of likely problems that may arise in respect either of cost issues or resource needs. They allow the management team to be proactive and to take the necessary pre-emptive actions to avoid likely problems. They can help management understand what changes are necessary if objectives are to be achieved. Models help the decision makers to understand the appropriateness of various policies or decisions. The consequence of a series of decisions or events can be evaluated and the longer term risk can be ascertained. 'What if?' modelling explains the likely impact of events or new policies and the levels of risk involved if changes to the business model are contemplated. It is not always necessary or appropriate to learn lessons through the hard experiences of failed implementation. The use of modelling facilitates the feedforward mode of cost control where existing performance can be challenged from an informed position and candidate areas for improvement identified and detailed plans developed. Traditional product costing systems are not the answer to management's decision needs. This level of precision is unnecessary. Managers need accurate information regarding product and service costs. Emphasis should be placed on understanding the behaviour of costs so that this may facilitate a better understanding of the business. The impact of all likely changes should be understood on the financial bottom line. This is not an intimidating task. The Pareto rule applies. Precise measurements of every cost element may not be necessary, and indeed some items may have very little impact on accuracy achieved.

3.7 PERFORMANCE MEASUREMENT

The third task of the management team relates to the ongoing control of the business. Performance measures provide an important role in policy formulation. They provide the milestones against which the business can measure its progress towards achieving its objectives or goals. They help signal if the plan is being implemented on target or if corrective action is needed. By pinpointing the areas needing management attention they serve to simplify the planning and control process, allowing managers or decision makers the opportunity to focus their attentions on just those few key variables that serve as the harbingers of good or poor performance. They provide managers with the opportunity to take corrective action in a timely (pre-emptive) fashion. A performance measurement system serves to direct behaviour within manufacturing. Behaviours and actions are motivated to conform to the performance targets. Unfortunately, the manufacturing response to some performance measures can be harmful for the overall good of the business. For example, a manufacturing business that is performing successfully against labour efficiency as its primary measure, could consequently experience higher inventories, longer lead times and increased levels of automation as well. In the effort to ensure that the business measures up to the selected targets, other aspects of the business performance get overlooked, hence the reason why an 'expect what you inspect' mentality has developed. While conformance to measures and targets might be achieved, the question of appropriateness emerges: **Are the measures being used the appropriate ones?**

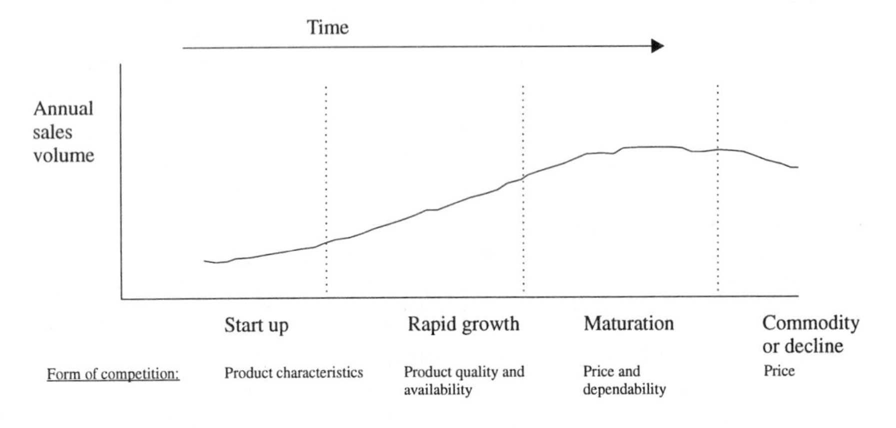

Figure 3.15 The product life cycle and changing order-winning criteria.

As the product moves through the various phases in its life cycle (Figure 3.15), the demands that it places on manufacturing will change. Changes occurring in the manufacturing environment, such as the introduction of new products, new processes, changing customer expectations, a new product technology or new business practices will necessitate a change in the variables that must be measured. A static performance measurement system that always focuses on the same measures will most likely provide incomplete and erroneous action messages. Yet actual practice in manufacturing has shown that performance measurement systems tend to be traditional in outlook, focusing mainly on the costs control aspects. A primary purpose of a performance measurement system is to guide company policy and decision making. The implications for decision making are obvious. The performance measurement system should be:

- flexible;
- easily understood by all;
- derived from the company's strategic objectives;
- dynamic.

The focus is on improving the effectiveness of the core value, adding activities and eliminating or reducing the need for non-value-adding work. For the decision maker (policy formulator) the key question is 'Are the actions and policies having the desired effects?' The performance measurement systems should provide sufficient guidance in the formation of tactical decisions. The preoccupation of traditional performance measurement systems with accounting information does little to help a company to understand why it has (or has not) made a profit. Indeed, the influence is often dysfunctional. An exclusive focus on costs might take the form of a company having measures in place to monitor:

- direct labour costs/variances;
- utilization levels;
- inventory turns;
- value-added level.

These measures, in turn, may result in the adoption of the following policies:

- long production runs to minimize set-up costs;
- minimization of the number of set-ups;
- out-sourcing to reduce the costs of direct labour;
- decoupling of demand and supply (using inventory) to achieve production economies;
- increased automation to reduce the cost of direct labour.

Although quite common, these policy reactions can result in negative consequences. Set-up costs can remain high, indirect support costs can increase, flexibility can be reduced, employee motivation can be minimized, the focus on quality can be reduced and the ability to bring new products to the marketplace can be reduced. The above scenario reflects the consequences of a company's preoccupation with one dimension of the strategy (cost perspective) to the detriment of the others, such as quality, predictability and flexibility. The performance measurement system should be adaptable, reflecting the changing requirements that occur throughout the life cycle of the product, facility or industry. Changes in the structure of the manufacturing environment can result in changes in the range and nature of the activities performed, the interdependencies of these activities, the resources they consume and the factors that drive their behaviour. A process choice decision to implement or phase over to a more highly automated process, displacing a previous need for labour, may necessitate that manufacturing should change its measurement focus from monitoring direct labour cost (traditional approach) to controlling the cost of the new technology solution and all its related support costs and benefits (such as engineering support, preventive maintenance, training, power, space, etc.).

The implications for decision making are many. If policies are to be successfully deployed within manufacturing, a broader perspective will have to be taken. The lessons to be learned from the emergence of ABC systems and recent writings in the area of performance measurement systems are that 'a firm cannot survive on secondary information' (measures based on cost) (Schoenberger 1986). The contribution and impact of policy decisions should be understood on all the dimensions of manufacturing strategy. To get this broader perspective requires examining their impact at the activity level within manufacturing. Manufacturing capability (and the resultant competitive advantage derived) is determined by the way resources are organized, controlled and used, not by the manner in which budgets are satisfied or cost controlled. The latter are merely a consequence of the former. According to Schoenberger 'The best strategy is doing things better and better in the trenches . . . The best leadership can come from having visible measures of what is going on in the trenches and on actions to achieve the desired rate of improvement.'

As well as helping to ensure that policies are successfully deployed throughout the organization, they also help to ensure that decisions and policies are formulated in a manner consistent to support the chosen manufacturing strategy. An **objectives network** can be created that explains the linkages between policies, critical success factors (CSFs) and performance measures. The network captures these cause and effect relationships and can help:

- identify the appropriate policies to be reviewed when critical success factors and performance targets change or when targets are not being met;
- explain the likely impact of specific policies;
- analyse and explain the trade off between policies and help find the appropriate 'fit' between different manufacturing policies.

The performance measures associated with the critical success factor provide targets and guidelines for policy formulation at lower levels in the decisional hierarchy that ultimately govern all the activities performed in the business. So, through performance measurement, policies can be adapted and the integrity of the business plan maintained.

3.8 POLICY MAKING

The fourth element of the four-task model is policy formulation and implementation.

Many manufacturing organizations often regard the flow and integrity of their information as restricting their competitive potential. Poor-quality information negatively affects the quality of manufacturing decision making. Perhaps one of the biggest flaws with manufacturing planning and control systems is that decision making is discouraged and expected from few. As the planning horizon being addressed reduces, the range of decisions addressed by the system reduces dramatically. The preoccupation of the system is with order numbers and schedules (Figure 3.16).

	Functions	Internal information	To manage engineering	To manage resources	To plan production	To manage materials	Internal information
Strategic	Business planning horizon		Manufacturing strategy formulation Manufacturing strategy implementation		Production plan formulation Production plan implementation	Manufacturing strategy formulation Manufacturing strategy implementation	Business planning monitor
Tactical	Master production schedule horizon				MPS generation MPS analysis		MPS monitor
	Requirement planning horizon				Requirements planning		
Operational	Shop floor control horizon				Scheduling Dispatching		Shop floor monitor

Figure 3.16 Narrowing the scope of decision making through MRP II.

This narrowing of focus has implications for policy deployment. Strategic policies do not serve as direct guidelines for the choices to be made at the tactical and operational levels by MRP II. Indeed, the master scheduling activities can operate quite comfortably without any strategic guidelines as inputs. Often strategic policies must get implemented and followed up outside of the system. Although MRP II plans and schedules the vast majority of manufacturing throughput, it does not explicitly address the associated strategic manufacturing choices that must also be made. The preoccupation of the system with order numbers and due dates serves to buffer the activities of the workforce from the broad requirements of customers and suppliers. In addition to being too narrowly focused, the planning process tends to be highly decoupled. The manufacturing planning architecture can be divided into three planning layers: strategic, tactical and operational. Each planning layer has its own view of the world. Sometimes this view is not well informed. Unfortunately, the only real integration that exists between the layers tends to be through the financial systems (Figure 3.17).

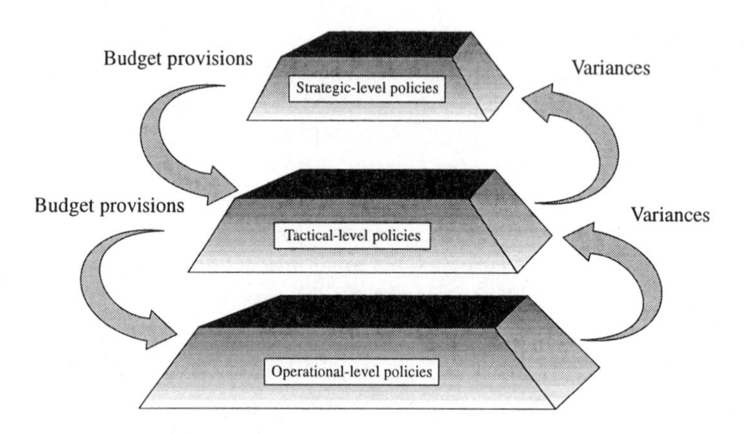

Figure 3.17 Integration through financial measures.

This data/information concentrates on consequences of manufacturing thus representing a limited means for integrating decision making. Decisions made at higher levels can often place unnecessary constraints on the manufacturing process. For example, a decision to freeze budget provisions in an effort to control costs may be detrimental to efforts aimed at improving manufacturing flexibility that have the potential for delivering sustained costs improvements. Although the logistics function is often identified as the candidate through which the business can differentiate itself from its competitors, conflicting objectives and manufacturing policies may render it ineffective.

3.8.1 THE DECISIONAL SYSTEM

Not all information processing is directly related to decision making. Indeed, decisions are not always obvious. Many decisions influencing the behaviour of the manufacturing system are taken by default. Nevertheless, whether explicit or not, they influence and control the behaviour of the manufacturing business and ultimately its performance. Decision making and policy formulation within manufacturing can be viewed as part of a three-layered model. The **decisional system** establishes the policies and rules that govern the **physical system** using the data provided by the **information system**. Within the decisional system, choices must be made at the strategic, tactical and operational levels. The range of strategic decisions can be classified into a number of policy decision categories or centres. Policies are selected within each category. Manufacturing business planning policies fall within the following six strategic decision centres (Figure 3.4).

- **Infrastructure strategy** explains how the business will be organized and controlled. Guidelines are given concerning the type of planning, inventory, quality systems to use and regarding the control and flow of information.
- **Make or buy strategy** determines which products or services the business will provide itself and those that it will out-source.
- **Human resource strategy** determines how the human resources will be organized and managed to meet the business objectives. The issues addressed include the design of the reward system and the determination of the skill sets required.
- **Facilities strategy** decides on the number of facilities and their locations as well as their business charters.
- **Capacity strategy** determines the response capability of the manufacturing facility. It provides guidance in relation to the forms and levels of capacity to invest in.
- **Process technology strategy** explains how the products will be manufactured or the services provided. Such issues as the types of process technology to employ and level of automation to be used are addressed.

Decisions taken in a decision centre should be coherent with each other and consistent with the overall business objectives. The decision to make or buy a product should support the firm's objectives regarding quality, predictability, flexibility and price. All decisions taken throughout the planning hierarchy should consistently support these aims. Decisions should not be made in isolation. Formulating manufacturing policies is like working with Lego. The policies must fit with each other as well as serving to shape the manufacturing business.

Decisions may result in conflicting messages at the lower level or in a conflict in resource availability/usage. Trade-offs should be made at the appropriate level. If the master planning function is unable to provide the level of flexibility required while supporting the guidelines set by the strategic capacity policy (with the level of resources and investment provisioned) then feedback should be provided to the manufacturing business planning function, which should be able to provide a resolution. Manufacturing objectives (demands) should be viewed from the context of all the decision centres. The task is to find the appropriate mix of policies that best serves to achieve the business's long-term vision. The process begins by identifying the business requirements such as the order-winning criteria and the business's current capability and performance. Each policy that impacts these objectives is reviewed in isolation to ensure that the current policy is still appropriate. A new policy may be formulated. For example, the decision may be taken to out-source a technology that was previously manufactured in-house. Next, this proposed policy is analysed in terms of its fit with the other strategic policies.

Any possible conflicts must be resolved before the agreed strategy can be adopted for implementation. Each decision centre provides a view or window to the manufacturing activities and the choice of tactics and policies that may be chosen. In viewing the manufacturing system through the make or buy decision centre, the implications of making or buying a product or technology can be examined. A business objective such as the improvement of quality can be influenced by the chosen make or buy and sourcing policies. The likely impact of policies on all the objectives must be examined. Existing policies must be evaluated and possibly new ones formulated. Strategies formulated and validated through the make or buy perspective will drive some of the policies and tactics that govern the manufacturing volume activities. There will also be a knock-on effect on the other policies and rules governing the manufacturing system.

3.8.2 THE DECISION PROCESS

Manufacturing policies are chosen in response to the customer-driven manufacturing vision. QFD can assist the strategic planning process, ensuring that policy directives are customer driven. The process begins with a set of loosely stated business objectives that represent the 'what' items that manufacturing must do in order to implement its long-term competitive vision. Implementing a 'pull' process and reducing manufacturing cycle time are possible goals or 'what' items. Next, manufacturing must identify 'how' each goal can be achieved. These 'how' items represent the manufacturing policies to be implemented. One or more policies may be identified to support each goal. For example, the implementation of statistical process control (SPC) methods **strongly influences** (i.e. enables) the implementation of pull processes within manufacturing. Each goal may be impacted by one or more 'how' items. Manufacturing management identify the likely influence each policy directive can have on each objective. This is captured in the QFD matrix (Figure 3.18). In addition to formulating the policy directives, manufacturing must also identify the targets or measures for each objective. The 'how much' items provide the milestones against which to measure the implementation of each directive. In the example, manufacturing has targeted December as the completion date for the decentralization of its quality department.

Implementing the manufacturing objectives using QFD is a multistage translation process. The agreed policy directives and measures at one level serve as the objectives or 'what' items at the next level of planning, where the process of determining policy directives and measures is repeated. In effect, both 'what' and 'how' items have targets associated with them. This decomposition process continues until all dependent policy directives are identified. At each planning level, the identified policy directives should be coherent and supportive of each other. In the example, adopting SPC methods before reducing the batch sizes is consistent. Often, the identified policy directives can be in conflict with each other. In Figure 3.19, the directive to reduce the level of in-house manufacturing can have a negative effect on manufacturing's ability to reduce its delivery lead time, especially if flexible suppliers are not available. The QFD matrix can be extended to capture the **fit** between policy directives. The triangular matrix above the 'how' items identifies whether the identified policies support each other or are in conflict. Two policy directives that are in conflict may be resolved by adjusting their measures. If the adjustment of the measures does not resolve the conflict, then the objectives or 'what' items may have to be changed. This implies a referral of the problem to the next higher level in the planning hierarchy.

The decision process can be further supported by enhancing the matrix to include competitive benchmark information. This can help to validate the quality of the objectives. The visual and structured quality of the QFD matrix supports a participative approach to policy formulation. Although the approach is structured, it does not guarantee that all the relevant policies directives likely to affect the manufacturing objectives will be considered. This structured decisional process is not always followed within manufacturing companies. Often, policies that are likely to affect the achievement of the business objectives are overlooked and are not formally reviewed or their impact analysed. Because of the diversity of possible manufacturing policy directives that can be made over time, an organizing framework (Hayes and Wheelright 1984) for policy categorization can be useful. The six decision

◎ Strong
○ Medium
△ Weak

Level 1

Objectives	Goals 'What?' \ Directives 'How?'	Implement SPC methods	Reduce batch sizes	Implement simultaneous engineering partnerships with suppliers	
Reduce delivery lead time	Implement pull processes	◎	◎		
	Introduce an enhanced flexible manufacturing system	◎	◎		
Reduce product cost	Reduce raw material costs	○	◎	◎	
	Reduce manufacturing cycle time	◎	◎		
	Reduce the level of finished goods		◎		

Level 2

Strategies 'What?' \ Directives 'How?'	Implement cross-functional training	Implement SPC training	Decentralize quality function	Establish cross-functional product development teams	
Implement SPC methods	△	◎	◎		
Reduce batch sizes	◎	◎	○	△	
Implement simultaneous engineering partnerships	△	△	△	◎	
Measures 'How much?'	December 1996		50% of shop floor trained		

Figure 3.18 Quality function deployment for strategic planning.

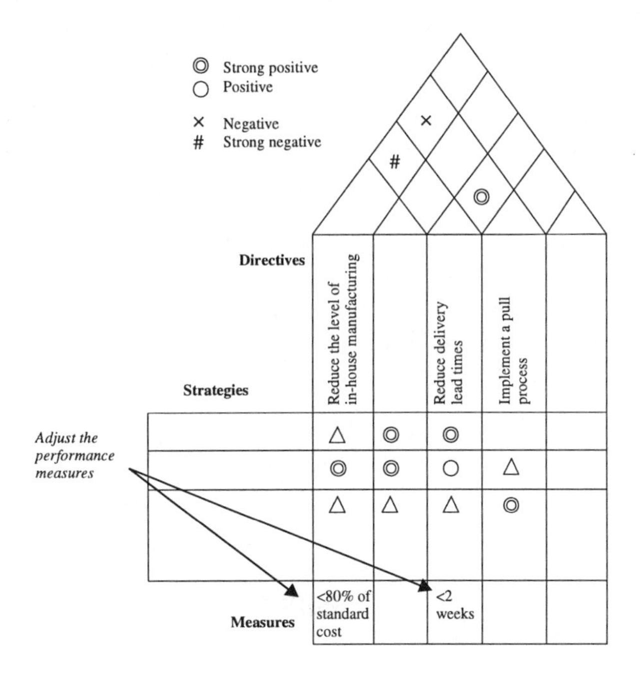

Figure 3.19 Conflicting policy directives.

categories outlined previously, provide a means for categorizing the possible set of policy decisions that can be made at the manufacturing business planning level. Each decision category is composed of the set of possible policy choices that can be made within that category. For example, a review of manufacturing's capacity policies could involve an evaluation of:

- the amount and form of equipment to be used;
- the amount and form of labour to be used;
- policies regarding overtime, shifts and utilization levels;
- policies regarding subcontract and alternative sources of supply;
- guidelines for raw material inventory, work-in-process inventory and finished goods inventory.

The QFD process for strategic planning can be further refined into an objectives network (Figure 3.20) that can identify the linkages between policies, critical success factors and performance measures. The CSFs are equivalent to the 'how' items in the QFD process.

The business objectives are translated into their critical success factors and key performance indicators (KPIs), informing the manufacturing business planning level of 'what' it must achieve. These 'what' items (including their measures) are reviewed against each decision category. For example, a review against the capacity policy category would involve an evaluation of the set of possible capacity policies or directives against each of the imposed key performance indicators. This evaluation involves a quantitative or qualitative analysis of the likely impact of various capacity policies on the KPIs. The analysis may be performed through quantitative 'what if?' modelling or through a knowledge rule base or both. The outputs from the analysis is a set of policy directives (how items) and measures (KPIs) that serve as the 'what' items for the master production scheduling level.

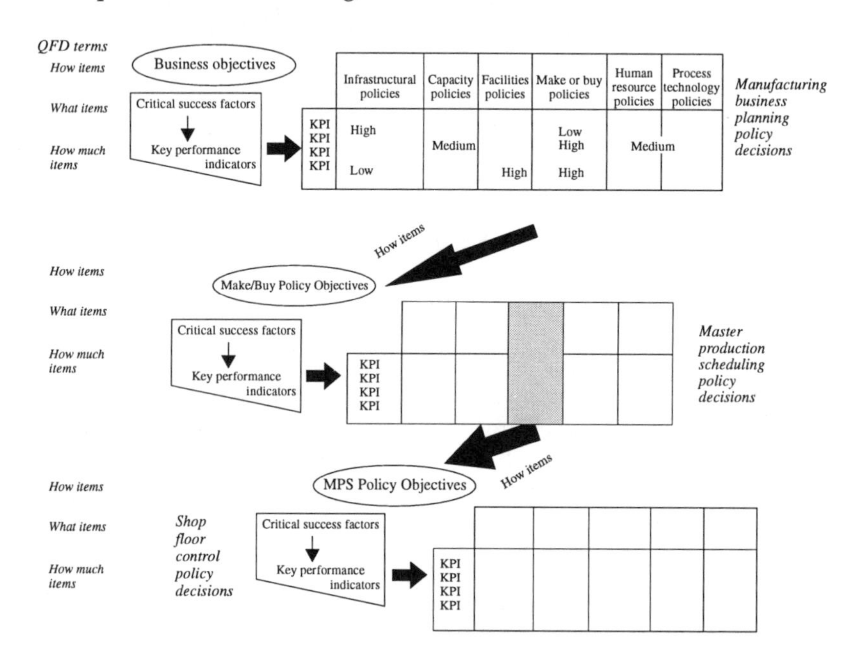

Figure 3.20 An objective network for policy formulation.

3.8.3 DECISION SUPPORT TOOLS FOR MANUFACTURING BUSINESS PLANNING

Information technology has a key role to play in educating and informing decision makers within manufacturing. In practice, policies tend be formulated on the basis of quantifiable factors such as the return on investment, likely throughput performance or cost figures. In these cases, the rules used to select the new policy are easy to understand and can be applied in an objective manner. The non-quantifiable or qualitative factors are not addressed in the same structured way. Often, they are not considered at all. The challenge for decision support tools for manufacturing planning is to provide the rules to the decision makers to explain the implications of both the quantifiable and qualitative factors in choosing the policies to run the manufacturing business. Digital Equipment developed a rule-based planning tool, as part of the IMPACS project, to help its manufacturing planners to choose the most appropriate make or buy policies for their computer manufacturing business in Galway, Ireland. The planning tool helps to explain the likely implications of different make or buy policy choices. Manufacturing's make or buy policies concern:

- the product technologies that it will make in-house and those that it will out-source;
- the process technologies to be employed;
- the activities it will carry out itself and those that it will out-source.

The impact of the make or buy policy should be evaluated with respect to the likely competitive position to which it will contribute. Rules are formulated to explain the implications of the make or buy choice on helping manufacturing to meet its business objectives. The following is a sample of the some business strategies that a firm could adopt within its industry. Some of the justifying factors supporting these choices are also included.

OBJECTIVE: INTEGRATE BACKWARDS IN THE SUPPLY CHAIN

Backwards integration implies manufacturing items that were previously purchased.

JUSTIFICATION

Backwards integration is justified if the market is subject to price variations, the risk of price increases is high, the continuity of supplies is threatened and purchase costs are excessive.

OBJECTIVE: ENSURE CONTINUITY OF SUPPLY

JUSTIFICATION

Entry into a new market and the emergence of new profit opportunities in new markets may cause concerns regarding the continuity of supply. Vendors may supply other competitors or perhaps enter the market themselves.

OBJECTIVE: REDUCE THE RISKS OF POSSIBLE TECHNOLOGICAL OBSOLESCENCE

JUSTIFICATION

A high rate of technology change within the industry, minimal adaptation of standards and high capital costs of the technology all contribute to the risk.

OBJECTIVE: PREVENT AGAINST THE THREAT OF NEW ENTRANTS IN THE INDUSTRY

JUSTIFICATION

If the cost of entry into the industry is significant (i.e. set-up capital, entrenched customer loyalty, etc.) then the likelihood of new entrants will be reduced.

OBJECTIVE: REDUCE CAPITAL REQUIREMENTS

JUSTIFICATION

A common ailment in many manufacturing enterprises today is overcomplexity caused by businesses becoming trapped into performing a wide a variety of activities. An unwelcome consequence of this is that capital costs and break-even points have most likely been increased, exposing the company to the vagaries of market demand fluctuations.

These cause and effect relationships can be reformulated into business rules that explain the implications of make or buy policies. In the sample rule base below, a planner can test the favourability of a make policy. Only a subset of the rules are depicted in the sample rule base. The '(....)' clause denotes the other conditions that must hold for the parent condition to be true.

TEST GOAL

RULE 10

IF (ensuring the continuity of supply) = 'High' AND
(the need to lower buying costs) = 'High' AND
(. . .
THEN (favourability of a make policy) = 'High'

RULE 12

IF (new entrants to the marketplace) = 'High' AND
(new markets emerging with high profits) = 'High' AND
(. . .
THEN (ensuring the continuity of supply) = 'High'

RULE. . .

QUESTION 1 (How would you define the level of new entrants to the market. i.e. High or Low?)

This proposition identifies rule 10 and tests all its clauses. In testing the first 'if' statement the inference engine backward chains to rule 12 and tests to see if it is true. In testing its first clause, the inference engine backward chains to question 1, which queries the user regarding the level of new entrants to the market. If the user indicates that the level of entry is 'High', the inference engine will then test the second clause in rule 12 and all subsequent clauses thus validating rule 12.

Control is then returned to rule 10. If the inference engine finds all the necessary conditions satisfied, then a make policy will be recommended. The user's response to the rule base describes the current situation within manufacturing and its marketplace. Different functional views can be input, resulting in different responses and sometimes a different policy recommendation. For example, a user may discount the importance of certain factors (rules or initial conditions). As a consequence these 'AND' clauses are given a lesser priority by the inference engine. The user can check why the recommendation was different and perhaps how production's view differed from marketing's perspective. If a policy recommendation is tested, the rule base will propose the necessary conditions that must be satisfied if the policy selection is to prove correct. The tool serves as a useful educational aid in identifying the cause and effect relationships associated with policy selection. This provides input to the QFD strategic planning process or the objectives network.

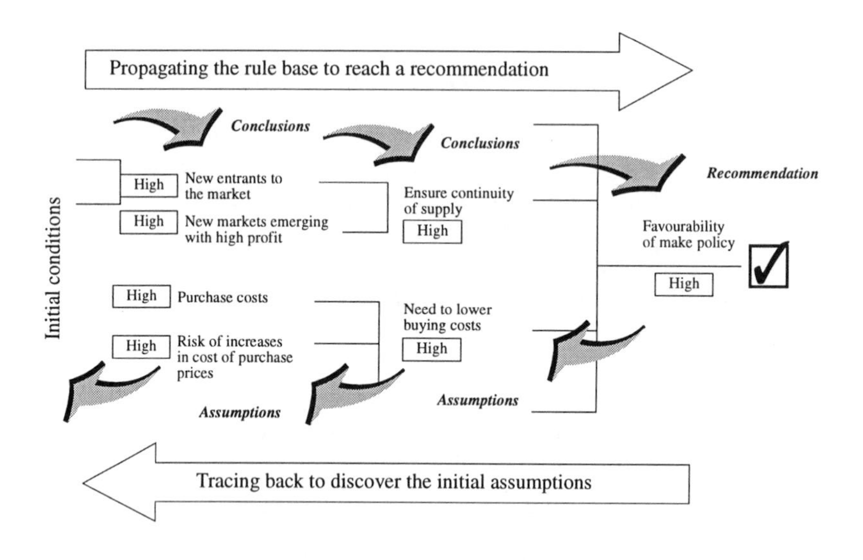

Figure 3.21 Sample rule base for the make or buy choice.

3.9 INTEGRATED PLANNING USING THE FOUR-TASK MODEL

Effective planning will not result from focusing on the tasks of resource planning, cost control, performance measurement and policy making in isolation from each other. The four-task model can be used to integrate these four perspectives through an integrated planning model. The business plans are validated at the appropriate level in the planning hierarchy before being deployed throughout the organization.

Manufacturing policies need to be reviewed periodically. The stimulus for review may be caused by changes in business objectives and their performance targets, changes to existing manufacturing policies or feedback through the performance measurement system indicating that existing objectives and their performance targets are unrealizable. First, a quantitative model of the total manufacturing environment is created. This is an activity based model that identifies the types and levels of resources used, resource drivers, performance capabilities, activity levels, costs and throughputs. Current policies and policy directives within each of the six manufacturing business categories are identified and their impacts on the manufacturing activity model is factored. This validated model that represents the current 'as is' situation is referred to as the 'baseplan'. The baseplan model is available to each decision category. The objectives network identifies the decision policy categories that are likely to have an influence on the impacted objective(s) and their measures. The planner or policy maker selects a policy category and the baseplan model (Figure 3.22).

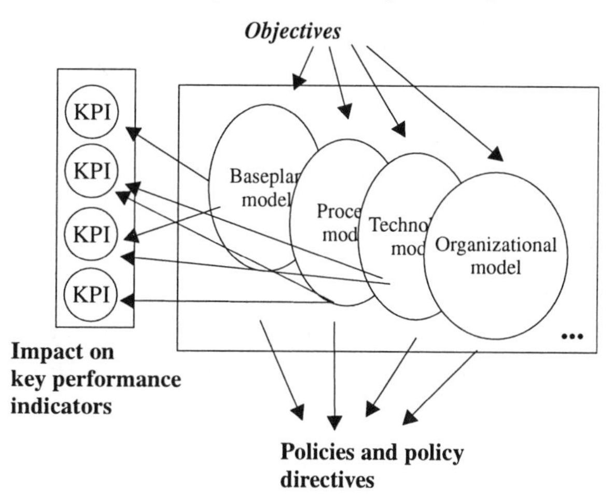

Figure 3.22 Policy decision making and key performance indicators.

Proposed policies and policy directives are factored into the baseplan model and a 'what if?' simulation is run to evaluate the likely impact of the changes on manufacturing's key performance indicators, resources, activity levels and bottom-line costs. The cumulative impact of a sequence of policy directives can be evaluated. Other additional models may also be constructed to complement the updated baseplan model. Like all of these models, their purpose is to explain the likely impact of policy changes on manufacturing's key performance indicators. These models could present the technology, organizational or process perspectives. When the analysis is complete, the objectives network is notified of the likely impact on the performance indicators. The network identifies the impacted manufacturing objectives and the other relevant policy decision categories that influence these. The process is repeated within each policy decision category until a balanced set of objectives, measures and policies have been established. This is validated at the long-term production planning level when the cumulative and aggregate impact of all policies and policy directives from the six policy decision categories are modelled. An overall activity, resource and cost model is created. Additional, more specific models focusing on skill sets, shifts, technologies, etc., may also be created.

The output from the business planning level to the MPS function is not the long-term production plan. Viewing the role of the master production schedule function as that of decomposing the aggregated plan would be overly constraining. Such a limiting role would result in an inefficient plan. The inputs to the master production scheduling level from the business planning level should include:

- objectives and performance targets;
- policies and policy directives;
- resource provisions and investment plans.

3.10 MAINTAINING BUSINESS MODELS

Modelling makes planning more effective. Nevertheless, traditional concerns regarding model maintenance must be addressed. It is often felt that a significant effort can be involved in developing and maintaining business models. These concerns are fuelled by a number of factors.

- Business models tend to be developed in isolation of each other. Adjustments made to one model do not automatically update the other relevant models.
- Business objectives are subject to change. This can imply that the purpose of existing models may no longer be valid.
- The manufacturing and business environments are subject to constant change.

These factors can increase the workload involved in keeping existing models up to date. A business should view the extra planning workload in the context of the downstream savings that can result. Effective upfront planning using QFD techniques has helped some companies, e.g. Aisin Warner, a subsidiary of Aisin Seiki Co. Ltd, Kariya, Japan, 'to halve their costs, double their productivity and quality in two thirds the time' (Eureka and Ryan 1988). Modelling helps the managers to increase their knowledge of their business and its processes.

Ensuring that models are **linked** together will simplify the maintenance effort involved in keeping business models valid. Objects in one model can be linked with objects in other models. For example, 'skill profiles' that may appear in the organization model could be linked to activities that appear in the process model, which in turn are linked to the resource and cost models. The interdependencies between the different views and models are numerous. Enterprise modelling tools can be used to create and manage the linkages between different models. Keeping all models linked should ensure that the implementation of business objectives will become more effective. Model **re-usability** should also help to reduce the maintenance effort involved. Re-using the 'baseplan' model to test various planning scenarios, minimizes the initial set-up effort required to validate the resource management, policy, performance and cost aspects of the proposed changes. The 'generic' rule bases created to assist policy making (i.e. make or buy) can be downloaded by a department. These generic models can then be refined and modified to meet their local planning needs.

These concepts of **linking** and **re-usable** components are (to some degree) also responsible for the enthusiasm associated with the object-oriented (OO) approach to software development. The application of these principles in OO software development promises:

- reduced solution development time;
- more consistent implementation of designs;
- less software maintenance effort.

Ironically, the origins of the object-oriented approach can be traced to the process modelling tools that were used within manufacturing in the 1960s and 1970s. Linking models together, integrating the tasks of policy making, cost control, resource management and performance measurement can help to improve the quality of policy deployment and reduce the maintenance effort involved in maintaining models and plans.

3.11 SUMMARY

Manufacturing management seeks a balance between the tasks of capability development and capability exploitation. The two tasks seem related to two modes of planning, i.e. top down and bottom up. Unless both tasks are done well, the business runs the risk of naively eroding its competitiveness. Increased competitive pressures imply that managers must incorporate both tasks as part of their ongoing planning process.

Capability exploitation is the hallmark of a management team that is well informed about the capabilities and behaviours of its manufacturing system. Tactics can be adapted to exploit current capabilities.

Further exploitation and extension of the capabilities of the manufacturing system become possible when bottom-up planning is effectively integrated into the overall strategic planning process. The four-task model is presented as a simple planning methodology and application that overcomes the biases with existing management control systems and empowers the management team with the information and explanations to accomplish both tasks successfully.

The role of product costing is questioned and the importance of work and resource behaviour is stressed. Decision making is differentiated from information management and the six manufacturing business planning decisions are presented. The four-task model accomplishes the tasks of validating the business plan and establishing guidelines relating to investments and resource provisions, one of the outputs being the long-term production plan.

The trade-offs inherent in decision making are highlighted through the use of an objective network in conjunction with decision support tools. Integrated planning throughout the MPC hierarchy is enabled through the use of an objectives network, which serves as the dynamic performance measurement system.

4 Master planning

4.1 INTRODUCTION

Chapter 3 has positioned manufacturing business planning as a planning activity with a strategic dimension. Business planning aims at determining the desired shape of the manufacturing system and at planning the transition towards it. Business planning is not an operational planning activity. It is not involved with managing the day-to-day response of the manufacturing organization to the market. Its outputs are formulated by means of policies. Those policies aim at guiding the migration towards the desired manufacturing shape (like policies to invest or disinvest), and determine the boundaries within which the operational planning activities should be executed.

This chapter covers master planning. Master planning is at the interface between manufacturing and the market. It is faced with the variability of demand, as well as with the degree of responsiveness of the manufacturing system. It couples the manufacturing system to its market and manages the actual response. Master planning might be considered as an **operational** planning activity, since it is concerned with **managing** the manufacturing system and the market on a continuous basis.

Many companies working with MRP-based manufacturing planning and control systems fully rely upon the material requirements logic itself to manage the response of the manufacturing organization to market demand. One of the primary purposes of this chapter is to draw the reader's attention away from the MRP core logic and towards the much more important concepts of master planning.

The function of master planning has become really important since markets evolved from supply markets to demand markets, as discussed in Chapter 1. Previous supply markets were constrained by production capabilities. Markets typically absorbed production supply. In current demand markets, manufacturers have to compete for market share and have to differentiate themselves from their competitors. Responsiveness and its management have become a competitive weapon, if not the most important one.

In the current competitive climate, master planning is becoming even more critical. A fundamental reason is with the market demand itself, which is becoming more difficult to manage, in at least three different respects. Firstly, market shares have become more volatile as a result of more intense competition. As a result, companies find more difficulty in assessing future demand. Secondly, competition has forced companies to try to meet customer expectations as best as possible, which has often led to extension of product offerings. Again as a result, the actual mix of product demand has become more difficult to anticipate. Thirdly, short delivery lead times have become a critical success factor. Commodity and fashion products need to be continuously available. Stock-outs result in lost sales and loss of market share. Delivery lead times for durable products, such as automobiles, hi-fi equipment, etc. have to be short if companies want to be in business. Shorter lead times means demands becoming firm at shorter notice, and again more difficulty in anticipating these demands.

In contrast, manufacturing companies have difficulty in responding. Manufacturing needs time to make available capacity, buy materials and produce the products before they can be made available to markets, if product demand has not changed meanwhile. Also, with current competition, there is a continuous focus on cost reduction. Stocks have to be reduced in size and value, yet responsiveness is mandatory.

Lead time reductions, flexible manufacturing techniques, just-in-time supply and manufacturing all address the same common issue of responsiveness to market demand. However, the results of these efforts remain below expectations without a successful planning approach.

In light of the above, master planning has become a key planning activity.

4.2 THE ROLES OF MASTER PLANNING

Master planning has two major roles as an interface between the manufacturing system and its market (Figure 4.1). These roles correspond with the two major flows of information between those two, i.e. a flow of information from the market through master planning to the manufacturing system, and another one in the reverse direction from the manufacturing system through master planning back to the market.

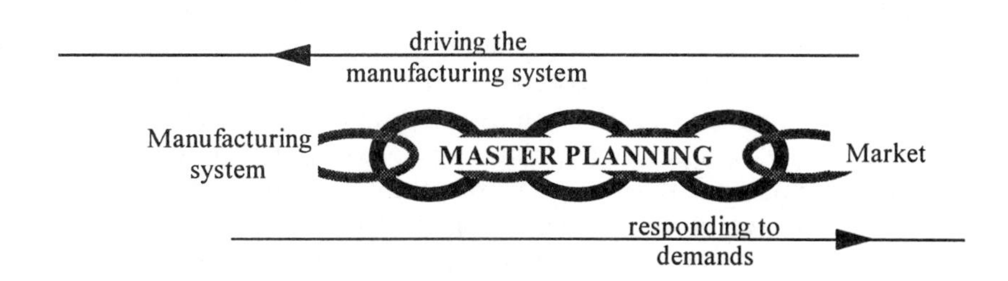

Figure 4.1 Master planning: interface between manufacturing system and its market.

The first role of master planning is one of synchronization. Master planning should ensure that the manufacturing system respond in a concerted and optimal way to the demands imposed on the manufacturing system. This first role involves the identification of the demands imposed on the manufacturing system and the generation and maintenance of a master schedule, a schedule of milestones to which all manufacturing activities should be geared. As such, master planning drives the manufacturing system in response to the demands, either firm or anticipated. Firm demands are demands that represent actual customer orders with promised delivery dates. Anticipated demands are demands that are expected to materialize in the near or distant future. Firm and anticipated demands together constitute forecasted demand.

In summary, the master schedule should be considered as the driver of all manufacturing activities. It expresses management's commitment in terms of an agreed response of the manufacturing system to forecasted demand. Therefore it should also be used as the basis for accepting new incoming demands. This constitutes the second role of master planning, which is to determine appropriate delivery dates by a process called order promising.

The two roles of master planning are covered by two separate but highly integrated functions, i.e. master scheduling and demand management (Figure 4.2). Master scheduling is about developing and maintaining the master schedule in response to firm and anticipated demands. Demand management attempts first to identify firm and forecast demand, via the processes of order entry and forecasting, and second to achieve order promising based upon the master schedule.

It should be quite clear to the reader that master scheduling and demand management cannot be considered in isolation from each other. Both are required and need to be highly integrated in order to realize an effective interface between the manufacturing system and the market.

Both roles also have to be performed in accordance with the guiding policies, as developed by business planning and discussed in Chapter 3. Guiding policies express strategic decisions to enhance the competitiveness of the manufacturing system, and that need to be taken into account by the operational planning activities. It is quite obvious that guiding policies may take very different forms, dependent upon the specific manufacturing environment. Some examples of guiding policies were given in Chapter 3, two of which are of utmost importance with respect to master planning. One is the capacity policy, the other is the responsiveness policy.

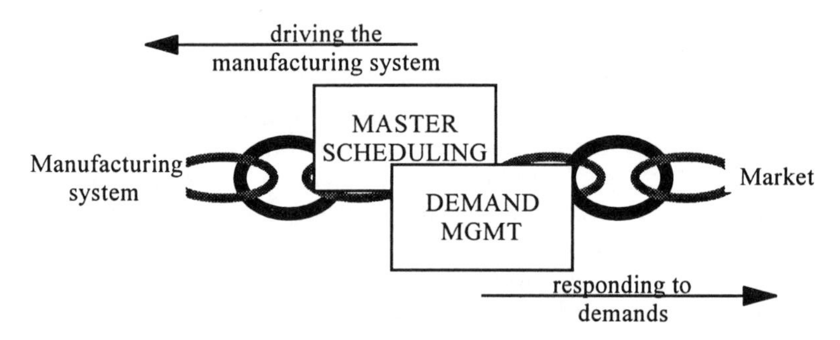

Figure 4.2 Functions of master planning.

The **capacity policy** is a statement of what the company is committed to in terms of investments or 'disinvestments' of capacity to manufacture. The capacity policy determines the boundaries of availability of manufacturing capacity. Development of a feasible master schedule implies that the master schedule be developed within these boundaries. The master schedule must not necessarily utilize the entirely available manufacturing capacity. It must simply be feasible in terms of its demands on the available manufacturing capacity.

The capacity policy may also define guidelines as regards subcontracting. Or it may determine maximum levels of investment in 'stored capacity', such as work-in-process or finished goods inventories. It may also determine target levels of investment in finished goods inventories as part of a level strategy in markets with seasonal demands.

The **responsiveness policy** is a statement of desired performance of response of the manufacturing system. It expresses how fast the manufacturing system is expected to respond to new and changed demands. As a result, the responsiveness policy is a reflection of or may determine the preferred manufacturing approach such as MTO, ATO or MTS.

Often, policies will express an objective within certain constraints, i.e. to maximize responsiveness for a maximum level of inventory investments, or vice versa to minimize inventory investment for a target service level. It is the role of master planning to achieve the objectives as expressed by the business policies. In this way it needs to act as an intelligent buffer that smoothes the impact of the variability of demand and allows the manufacturing system to respond within some predefined lead times. Demand smoothing is typically achieved through planning of safety inventories as part of the master schedule.

Most companies do not yet make explicit the distinction between the business planning level and its accompanying guiding policies on the one hand and the operational (day-to-day or week-to-week) planning activities on the other. Explicit distinction is, however, required in order to separate the business planning activities that are discrete in nature and serve to **shape** the manufacturing system from the operational planning activities that are continuous in nature and serve to **run** the business according to the established guidelines. The authors believe that the competitiveness of a manufacturing system can be increased by clearly separating the shaping activities from the day-to-day managing activities. The more specific guidelines that can be developed, the more operational planning is facilitated. This actually calls for concentrating the shaping decisions as much as possible within the business planning layer, and 'reducing' the operational planning activities to day-to-day managing activities within clearly established guidelines. This recommendation is not in conflict with the ideas to empower decentral organizations or devolve decision making. Business planning could or should be supported by as many people from the organization as possible. The recommendations simply want to highlight the need for carrying out business planning functions at **discrete** intervals in time, and not continuously (as part of the master planning activity).

As such, two important objectives are achieved:

- Business planning is formalized as a shaping activity. Thought is given to developing policies to optimize performance of the business. Formal policy development helps clarify and communicate the policies across the entire organization;
- Operational planning activities can be tuned as lean activities to perform their managing role within the guidelines of the established business planning policies. Of course, some 'policy' decisions may still have to be taken within each of the operational activities. But these policies should be local in nature with minimum impact on other activities.

This book assumes that manufacturing as a term covers a broader spectrum of activities than only production. Manufacturing companies may carry out activities ranging from engineering through production, distribution and installation up to on-site test of their products. However, many – if not most – manufacturing companies have a predominant focus on production. Additionally, most production companies have an MRP-based planning and control architecture.

The discussion on master planning for such companies is the subject of the first and major part of this chapter. The extensive discussion also includes an evaluation with respect to the concepts of JIT and synchronous manufacturing. The discussion should make clear that MRP II as a planning framework is **not** ideal across all types of industry. The follow-on part is focusing on the search for an alternative planning approach, suitable for those companies for which MRP II does not prove to work. The chapter ends with what should be one of the major credits of this chapter, which is to map planning concepts to types of industry according to functional fit and actual industry requirements.

The treatise on MRP II in the first and major part of this chapter is rather technical but necessary to really understand its premises, strengths and weaknesses. This part serves as lead-in for a more conceptual part, which tries to summarize one of the most interesting areas for further developments in planning and control, which is the development of appropriate planning and control systems for process and semiprocess industries.

4.3 MASTER PLANNING IN THE MRP II ENVIRONMENT

This section extensively discusses the MRP II planning approach, and in particular the key role master planning plays within that planning framework. It starts with some necessary introductory sections aimed at:

- positioning the terms 'master production scheduling' and 'material requirements planning' (section 4.3.1);
- grasping existing practice within the domain (section 4.3.2);
- clarifying some terminology as a lead-in into follow-on sections (section 4.3.3);
- discussing some essential 'shaping' activities with respect to master planning (section 4.3.4).

The two roles of master planning are subsequently extensively discussed: firstly its role as driver of the manufacturing system (section 4.3.5) and secondly as basis for order promising (section 4.3.6). Subsequently, the MRP II-based architectures are discussed for MTS and

ATO environments (section 4.3.7). Finally, there is an overall evaluation of the MRP II planning framework (section 4.3.8). Firstly, the applicability of MRP II is being evaluated. Secondly, MRP II is contrasted with the basic concepts of recent philosophies such as JIT and synchronous manufacturing. The extension from production planning towards enterprise planning is briefly discussed (section 4.3.9).

4.3.1 INTRODUCTION

The purpose of this section is to discuss master planning for production companies that have adopted the MRP II planning and control architecture. Master planning for these companies conforms with the generic objectives and model as introduced in the previous section. As indicated earlier, it consists of two separate but highly integrated functions: master scheduling and demand management. However, master scheduling will be called master production scheduling, thereby referring to the production nature of the companies currently under consideration.

Master production scheduling actually originated as a term within the well-known MRP II production planning and control architecture. This architecture is centred around the material requirements planning core module (see also Chapter 2). MRP is a planning approach that recognizes the concept of dependency of demands within production environments. The demand for any item that is required to produce a parent item can be derived from the production schedule of the latter item. The concept of dependency of demands has been extended to a quite powerful calculation logic to plan replenishments for each of the raw materials, components and subassemblies within a production environment. This logic looks at the demands for a certain item, derives net demand by taking account of on-hand availability and open orders (**netting**), and plans replenishments according to replenishment policies (such as **lot sizing**) and safety parameters. The planned release dates for these replenishment orders are offset from planned due dates by the purchase or manufacturing lead time of the concerned item (**lead time offsetting**). Planned replenishments for manufactured items result in turn in demands for component items (**explosion**).

This netting, lead time offsetting and explosion logic is a top-down calculation process that needs a triggering input. This input is the MPS. It is expressed as an anticipated build schedule. The MPS items may be finished products in an MTS environment or standard product modules in an ATO environment. The MPS drives the MRP logic. The logic ensures that all of the production and procurement operations are synchronized in support of the MPS.

4.3.2 EXISTING PRACTICE

The IMPACS project, as referenced before, has focused part of its attention to master production scheduling and has undertaken a review of existing practice in the MPS domain, in particular by means of a questionnaire and a set of European-wide interviews.

The major finding was that current master production scheduling approaches and practices were felt to be unsatisfactory by most companies. In the following text, findings will be categorized in two sections: findings that are related to the role of the MPS as driver of MRP and those that are more related with the role of the MPS as basis for order promising.

4.3.2.1 Role of the MPS as driver of MRP

Current practice in terms of the MPS as driver of MRP is often found to be weak or even poor in three different respects. Firstly, the MPS is not used as the major tool to manage responsiveness as it actually should be. Secondly, its development approach clearly suffers from the lack of control with respect to bottleneck resources at the levels of aggregation at which master production scheduling typically operates. Thirdly, the lack of 'good' integration with the business planning level is also a major theme in criticism. These three themes are successively discussed in more detail.

Lack of management of responsiveness

The first wave of MRP implementations after the formal introduction of MRP by Orlicky (1974) has been led by the capital goods and durables industry segments, which were faced with a large materials management problem. Although the need for and role of master production scheduling was clearly highlighted by Orlicky in 1974, many of these companies started off by implementing the core material requirements planning logic and simply driving it by a statement of forecasted demand, thereby bypassing the MPS function.

It is clear that those companies suffered severely from the adopted approach. As soon as actual demand deviated from forecast demand, companies were faced with the cumulative lead times of procurement and production when trying to adapt their purchasing and production operations to the changed demands. The production organization continuously relied on expediting to make the impossible happen. Quite frequently, parts or subassemblies were 'stolen' to make possible a sudden but urgent shipment, while delaying other shipments for which those parts or subassemblies had been planned.

In order to become less vulnerable to unexpected changes in demand, companies planned all kinds of safety policies in procurement and production. Safety policies included inflated lead times for procurement or production and high safety stocks for critical materials. Safety parameters were usually determined on the basis of Pareto analyses (also named ABC analyses), which aimed at separating the critical A class items from the respectively larger number of less critical C class items. Categorization criteria included annual volume, supplier or process reliability and lead time. Safety parameters were set accordingly, at quite high levels for critical items and lower levels for less critical items.

These safety policies were planned to cover a variety of uncertainties, including on-time delivery by suppliers, conformance to quality and demand uncertainty. The uncertainties can broadly be categorized into supply-related uncertainties and demand-related uncertainties. Supply-related uncertainties are those that relate to the supply of the concerned item ('Will the planned numbers of a certain item be produced or supplied on time while conforming to quality specifications?'). Demand-related uncertainties are those that relate to the demands for the concerned item ('To what extent are we sure of the expected time-phased requirements?'). Despite the different nature of the demand- and supply-related uncertainties, those were typically addressed using the same safety policies, planned by MRP by means of corresponding safety parameters.

Only much later did most companies realize the importance of the MPS function as a kind of buffer between the production system and the variability of product demand. Distinction was made between the statement of expected demands, and the MPS as the 'agreed' response or anticipated build schedule. This anticipated build schedule is expected to optimize the response of the production organization (e.g. in terms of load on production resources), but may additionally include some 'overplanning' to be able to respond to unexpected changes in demand. In addition it ensures more stability by protecting the production system from unexpected changes in demand.

However, with the introduction of the MPS as a kind of buffer between the manufacturing system and its market, most companies forgot to remove part of the safety policies planned throughout the value chain. They are now faced with a combination of safety policies within procurement, production, master production scheduling and hence a corresponding excessive cost to provide for a desired degree of responsiveness.

The safety parameters within MRP may be justified and can be tolerated in the following two cases. If the item is subject to a perceived form of supply related uncertainty, safety may have to be planned to protect the company from non-conformance to quality specifications, late deliveries, etc. Secondly, if the item is used as a spare part, part of the item demands are independent demands and the company may want to provide for a certain service level by planning a certain safety stock.

These safety policies, planned for within MRP, are not visible at the level of master planning. The result is extra safety stock or work in process, which can only be traced by querying the information for the concerned item within MRP. Secondly, they are typically the result of actions within procurement, production planning, etc. without regard to the specific market uncertainties at the level of finished products with which the company is faced.

The MPS, in contrast, provides the ideal vehicle to protect the production system against market uncertainties. Any overplanning is clearly visible and results – via the MRP explosion and planning logic – in matched stocks at component and subassembly level. However, few companies use the MPS as the prime vehicle to protect themselves against demand variability.

Perceived weaknesses of the development approach

The MPS should in the first place be a feasible schedule and be developed within the boundaries of available production capacity. The feasibility of the MPS is typically assessed using rough-cut capacity planning techniques. Rough-cut capacity planning (RCCP) consists of a rough-cut simulation of the anticipated load of a tentative MPS on production resources. This technique is however felt to be unsatisfactory to ensure the proper identification of temporary overloads on production resources. As a result, companies develop their master production schedules with much experience and gut feeling, without much quantitative analysis. MRP is subsequently run without a sufficient degree of certainty as regards the feasibility of the overall production programme.

It is only through capacity requirements planning (CRP), a deterministic simulation of capacity requirements based upon the actual output of MRP, that a more detailed assessment of feasibility can be performed. In case of minor resource overloads, the output is used to adjust the availability of capacity (e.g. number of work shifts, overtime, subcontracting, etc.). If the production programme appears to be unfeasible, the MPS needs to be reassessed.

Unfortunately, CRP fails as a good decision support system. Firstly, its output can only be made available after the MRP run, which for most companies takes considerable time. (Note that this hampers the use of CRP as a what-if simulation tool during development of the MPS.) Secondly, the CRP output identifies overloads but does not provide decision support as to how these overloads could best be avoided.

The standard MRP II approach consisting of a first stage of MPS development and a second stage of MRP/CRP calculations clearly shows weaknesses within manufacturing environments where capacity constraints are an important issue. The MRP II limitations are typically compensated using the experience of good master schedulers. Companies that fail to recognize the weaknesses of the MRP II approach end up with either a total failure or very mediocre results.

Lack of 'good' integration with business planning

The traditional view on integration between business planning and master planning is through the production plan. The production plan is considered as the output of the sales and operations planning activity, a subactivity within business planning. APICS defines sales and operations planning as:

> the function of setting the overall level of manufacturing output . . . to best satisfy the current planned levels of sales (sales plan and/or forecasts), while meeting general business objectives . . . ; one of its primary purposes is to establish production rates that will achieve management's objective of maintaining, raising or lowering inventories or backlogs, while usually attempting to keep the workforce relatively stable.

The output is expressed over a medium- to long-term horizon and is usually stated as the target monthly production output by product family. As the definition already indicates, the production plan is especially relevant in MTS businesses with seasonal sales, where a balance must be found between overall sales expectations and production capabilities. This balance often results in a level strategy where expected variations in sales are covered by planned increases and decreases of inventory levels and a more stable production plan. (Since it is essential to be able to evaluate the production plan in terms of its loads on production resources, it is important that product families be defined from a production perspective, i.e. as groups of products sharing similar production resources. Products could be grouped differently for sales or marketing purposes.)

The production plan serves as a constraint for development of the MPS. The master production schedules for the individual products are expected to be a decomposition of the production plan (e.g. from monthly figures by product family for the production plan to weekly figures by product for the individual master production schedules). Or vice versa, the master production schedules for the individual products should aggregate back to the predefined production plan.

Attention must be paid to ensuring that the MPS development process is not overly constrained by imposing a certain production plan, which – after all – might be reviewed only once per month during a business planning review meeting. The MPS should always be seen as the **prime** vehicle to manage the response of the production system to the market. Care should be taken to guarantee that the MPS can respond quickly to perceived demand changes.

In other companies, production planning and master production scheduling are one and the same process, which ideally should be guided by means of guidelines as established by business planning. Integration between production planning and master production scheduling ensures that responsibility is clearly localized and results in a responsive master planning process. Integration does not imply the absence of quantitative evaluation at manufacturing business planning level. On the contrary, tentative master production schedules could even be developed for 'what if?' simulation purposes during development of the manufacturing business plan. These simulations should, however, serve the development of appropriate policies, which should be considered the **real** output of business planning. Master production scheduling should not feel constrained by a production plan that was developed for quantification of a certain business planning scenario. Rather, it should work within the boundaries of the agreed policies continuously to optimize the response and the responsiveness of the production system to market demand. Figure 4.3 illustrates the recommended view versus the traditional view with respect to the guiding inputs for MPS development. Much more important than the production plan itself are the business policies that should guide the development of the MPS in response to typically highly dynamic markets.

The change is mandatory to meet the objectives of responsiveness in the current competitive environment. The development of an MPS within the boundaries of a predefined production plan dates from a period when markets evolved only gradually and responsiveness was less of an issue.

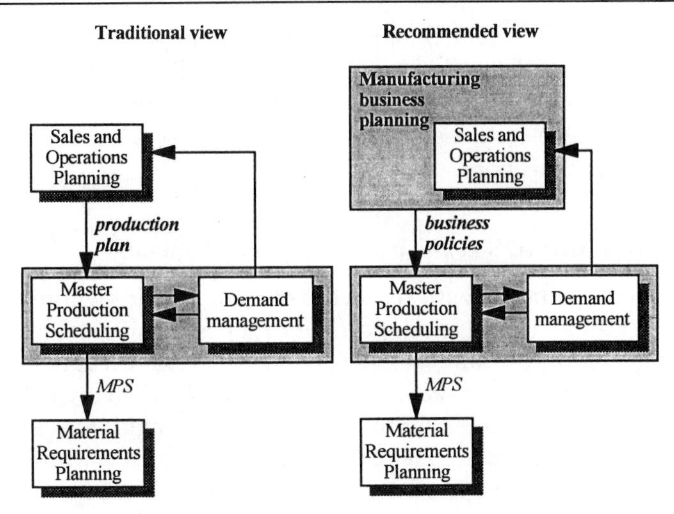

Figure 4.3 Traditional view versus recommended view as regards integration between 'master planning' and 'sales and operations planning'.

In general, the need for and use of a separate production plan must be carefully evaluated, since it may act as an unnecessary constraint on the MPS development process. A production plan, if any, could even be summarized as a set of time-phased target inventories by product family and be one of the guiding policies. In this way, the time-phased production plan figures no longer constitute a constraint on development of the MPS. The objectives to achieve are expressed by the inventory targets.

Guiding policies are important when it comes to integrating business planning and master production scheduling. The identified priorities become highly visible and can be used as the basis for identifying performance measures. Performance measures are required to evaluate the performance of a certain MPS or a series of consecutive releases of master production schedules over time (e.g. stability of the MPS, responsiveness to demand). Comparison of master production schedules against predefined production plans may be too rigid. Much more important are customer service level, flexibility of response and adherence to predefined inventory targets, if relevant.

The non-existence of clear policies often results in vagueness of customer service requirements and expectations as regards responsiveness of the production organization. Clear policies trigger an overall evaluation of the responsiveness of the production organization, which in turn may identify opportunities for better master planning. One such opportunity is in the recognition of commonality among products, which in many cases may lead to a significant reduction of the cost of providing a certain degree of responsiveness. This will be further discussed in later sections.

4.3.2.2 Role of the MPS as the basis for order promising

Another major criticism with respect to current practice within the MPS domain relates to its role to support the order promising process. Order promising stands for all activities related to the acceptance of an incoming customer order against a certain (feasible) delivery date. Order promising in an MTS environment is rather straightforward, since promising can be done against available stock in finished goods warehouses. The order promising function in other environments (ATO and MTO) is much more complex and requires good integration between the MPS function and demand management. Ideally, the MPS should be used as the basis for order promising. It expresses the response of the production system to perceived demand. When taking account of promised customer orders, 'available-to-promise' information should be derived and used as the basis for further promising. Many companies have only a poor integration between the MPS function on the one hand and demand management on the other. Customer orders are typically promised against some average perceived feasible output, without checking projected availability of finished products, or required product modules. One of the immediate results is the frequent changes of an MPS even in a short time horizon, resulting in a high level of expediting activity. But even more important is the weak customer service level that many of these companies prove to achieve. A higher customer service level at even lower cost can be achieved by really using the MPS as a vehicle for customer order promising.

The review of current master production scheduling practice, executed within the IMPACS project, has clearly revealed a great need for improvement. Pertinent questions are as follows:

- How could the MPS effectively be used to manage responsiveness?
- How could the effectiveness of the MPS development approach be improved?
- How should a reliable basis for order promising be implemented?

The vision for master planning embraces a highly integrated system of two functions, master production scheduling and demand management, aiming at:

- driving the manufacturing system in accordance with the guiding policies from business planning;
- responding to demands within the capabilities of the production organization.

The following discussion uses many specific examples. In order to create a common basis for understanding, some terms are first explained in the next section.

4.3.3 TERMINOLOGY

4.3.3.1 Planning table

The MPS is in the first place the driver of material requirements planning. As such, it should be considered as the anticipated statement of 'production'. The products or product modules, in terms of which this anticipated statement is expressed, are called MPS items. The standard representation of the MPS for each of its items is very similar to the time-phased representation of MRP records. Figure 4.4 illustrates an MPS for a certain item X. The information in Figure 4.4 could be called a planning table for MPS item X.

MPS item X	week 1	week 2	week 3	week 4	week 5
Forecast	10	10	8	8	10
Customer orders	4	4	0	0	0
Net demand	10	10	8	8	10
Master Production Schedule	10	10	10	10	10
Projected Available Inventory	2	2	4	6	6
Scheduled receipts	10				
Pl. order releases	10	10	10	10	

On hand = 2, production lead time = 1 week

Figure 4.4 Standard representation of the MPS record.

4.3.3.2 Time buckets

The planning table is a time-phased representation of information for different planning entries. Time phasing is typically done using a series of time buckets, which segment the entire planning horizon. Time buckets represent a certain period of time. They are used in the first place to accommodate easy visualization of the underlying information.

Early MRP systems were typically bucketed. Requirements and open orders are consolidated by time bucket, and new orders are planned by time bucket. For such 'bucketed' MRP systems, it is standard practice to have the time buckets of MPS planning tables equal to those of the MRP system. Buckets are typically 1 week in length across the entire planning horizon.

Nowadays MRP systems have a bucketless approach. Requirements and released orders are dated and considered individually, rather than consolidated. New orders are planned according to the MRP policies and parameters, but without making reference to bucket sizes. For these MRP systems, time buckets for MPS planning tables only serve to simplify the visualization of information. Often, time bucket sizes are weeks at the near horizon, and months or even quarters at the distant or far horizon.

In Figure 4.4, five time buckets are considered, each 1 week in length. The actual planning horizon may continue beyond time bucket 5, and should at least cover the cumulative procurement and production lead time for the item X. This minimum planning horizon is required to ensure that all of the components and subassemblies can be made available in accordance with the MPS.

4.3.3.3 Planning rows

The planning table consists of different entries or planning rows.

Total expected demand is represented by the entry **forecast**. Note that forecasts refer to total forecast, or forecast including firm demands. Firm demands are those that correspond with firm customer orders. Anticipated demands are demands that are expected to materialize in the near or distant future. Some of the expected demand has already been confirmed by customers and promised by the production company. This firm demand is represented by the entry **customer orders**. In Figure 4.4, existing customer orders amount to 4 units of MPS item X in weeks 1 and 2. The firm and anticipated demands are the demands imposed on the production company. These demands may emanate from external customers or internal customers, such as distribution centres or the departments owning the finished goods stocks. **Net demand** reflects the demands that the company will use as a basis for developing the MPS. In Figure 4.4, net demand simply repeats total expected demand, as indicated in the entry **forecast**. However, the calculation of net demand is actually subject to the forecast consumption rules, which will be discussed as part of the discussion on order promising.

The entry **master production schedule** states the completion of units by time bucket to meet firm and anticipated demands. The MPS is a statement of planned production and must be distinguished from the statement of demand itself. The existing MPS together with the available on hand of 2 units results in a projected available inventory profile (PAI) as represented by the entry **projected available inventory**. The PAI identifies what is expected to be available after taking into account production and consumption by net demand. The PAI figure for week 3 is for instance calculated as the PAI figure of week 2, plus the production of 10 units, less the net demand consumption of 8 units.

The entry **scheduled receipts** identifies the open orders, i.e. those orders released into production and most likely to be in progress. The orders are dated by the planned completion dates. The entry **planned order releases** identifies those orders not yet released into production. These orders are dated by the planned release dates. The planned release dates are offset from the planned due dates by the production lead time for the item X. The orders for the MPS items are not **automatically** planned by MRP. An initial or tentative MPS may be generated automatically using MRP logic as a starting point. Final development of the MPS is a highly critical process that is under the responsibility of the master production scheduler. Within the nearby horizon, the concept of firm planned orders is used to prevent MRP from changing these orders at the next MRP run. MRP will trigger production planners to release the orders in the action time buckets (typically the first time bucket, i.e. week 1).

4.3.3.4 Time conventions

With visualization by time bucket, time conventions are important. Total expected demand as represented by the entry **forecast** is typically not dated. The forecast entry simply represents what is expected to be consumed **during** each individual time bucket. Customer orders, on the other hand, are dated. The numbers by time bucket are consolidations of requirements for individual customer orders, which have specific delivery dates associated with them. **Net demand** derives from firm and anticipated demands. Net demand is typically represented by time bucket, although it partially consists of firm customer orders with specific delivery dates. Net demands are to be met by the MPS. A critical question relates to how and when the MPS is expected to meet the net demands or 'How should the precision of MPS planning be adjusted to the precision of customer order promising and shipping?'.

One of the possible time conventions is to consider that all net demands should be satisfied by the end of the corresponding time bucket. This actually implies that the numbers in the **MPS** entry be dated according to the end of the corresponding time bucket, and not the start or middle, as would equally be possible. Assume that the time buckets 1–5 are consecutive weeks, starting 2 January 95. The 5 (work) weeks end 6 January, 13 January, etc. up to 3 February. Assume that the firm demands of 4 units each are dated 6 January 94 and 13 January 95. Anticipated demands as well as MPS orders are dated according to week ends. A bucketless visualization of the planning table would look like Figure 4.5. At first sight, there is no difference between bucketed and bucketless information. This changes however if orders were not promised exactly at the ends of the time buckets. In Figure 4.6, firm demands are dated respectively 3 January and 11 January. The PAI figure first decreases to −2 units (2 units on hand, minus consumption of 4 units), and subsequently increases to 2 units (as a result of production and consumption by anticipated demand, expected to occur simultaneously at the end of week 1). The bucketless visualization identifies a problem situation. The firm demands cannot be delivered as promised.

Date	Customer orders	Anticipated demands	MPS	PAI
On hand				2
8 Jan	4	6	10	2
15 Jan	4	6	10	2
22 Jan		8	10	4
29 Jan		8	10	6
5 Feb		10	10	6

Figure 4.5 Bucketless representation of Figure 4.4 – example 1.

This problem situation is typically overcome in one of the following ways. All approaches aim at compensating for the inaccuracies caused by visualizing and scheduling by time bucket. A first approach is to use safety stock to compensate for planned shipments preceding scheduled production. A second approach is to use safety lead time to offset production from planned shipments (e.g. by 1 week). Another possibility is to revisit the time convention of the MPS orders and to schedule completion of the MPS orders at the start of the corresponding time buckets (Figure 4.7). In doing so, production is scheduled to anticipate actual demands by roughly half of the bucket size. The smaller the bucket size, the less significant the temporary stock build-up and the less expensive the approach will prove to be.

Date	Customer orders	Anticipated demands	MPS	PAI
On hand				2
5 Jan	4			-2
8 Jan		6	10	2
12 Jan	4			-2
15 Jan		6	10	2
22 Jan		8	10	4
29 Jan		8	10	6
5 Feb		10	10	6

Figure 4.6 Bucketless representation of Figure 4.4 – example 2.

Date	Customer orders	Anticipated demands	MPS	PAI
On hand				2
2 Jan			10	12
3 Jan	4			8
6 Jan		6		2
9 Jan			10	12
11 Jan	4			8
13 Jan		6		2
16 Jan			10	12
20 Jan		8		4
23 Jan			10	14
27 Jan		8		6
30 Jan			10	16
3 Feb		10		6

Figure 4.7 Bucketless representation of Figure 4.4 – example 3.

4.3.3.5 Rolling planning horizon

The MPS is not a static statement of anticipated production. Rather, it has to dynamically respond to changing product demand, while taking account of deviations between actual production and anticipated production as well as of changes in planned available capacity. All of these changes occur while time progresses, with planning horizons rolling through time.

MPS item X	past due	week 2	week 3	week 4	week 5	week 6
Forecast		10	8	8	10	?
Customer orders	1	8	4	0	0	0
Net demand		10	8	8	10	?
MPS	1	10	10	10	10	?
PAI		0	2	4	4	?
Scheduled receipts	1	10				
Pl. order releases		10	10	10	?	

On hand = 0, production lead time = 1 week

Figure 4.8 Planning table of Figure 4.4 after 1 week.

In Figure 4.8, the planning table of Figure 4.4 is illustrated after 1 week. The planning horizon extends into the future, starting from week 2. An additional column, **past due**, has been added to track firm demands that have not been fulfilled, as well as planned production that has not been realized as anticipated. In the example of Figure 4.8, actual orders for week 1 have turned out to be 12 units, in contrast with the original forecast of 10 units. Ten units were scheduled to be completed by the end of week 1. Only nine were completed, one unit is past due. As a result, not all demands have been satisfied by the end of week 1. The on hand of 2 units, incremented with completion of 9 units, resulted in availability of 11 units, against a demand of 12 units. The **customer orders** entry has therefore 1 unit past due, which still has to be satisfied. Extra orders have arrived for time buckets 2 and 3, which – within the example – does not result in a change of net demand. Yet, with modified starting conditions at the start of the planning horizon, the PAI profile has changed from that in Figure 4.4. The entire profile has actually been reduced by 2 units, as a result of extra demands in week 1. This phenomenon may be a sufficient reason to revise forecasts and/or the MPS, which has to respond accordingly. The extent to which the MPS can be changed is, however, limited by constrained materials and capacity availability. Changes to an MPS must be verified against availability of materials and capacity. The issues related to development of an MPS will be discussed later in this chapter.

The planning horizon for an MPS item should cover the cumulative lead time for procurement of components and subsequent manufacturing steps. As the planning horizon rolls through time, new **forecast** and **MPS** numbers have to be decided upon.

Earlier it was mentioned that not all time buckets need to be the same in length. With current bucketless MRP systems, the typical approach is to go for small time buckets over the short horizon and larger time buckets over the long horizon. Small time buckets provide more visibility and accuracy, but require more planning effort. Using larger time buckets allows the planning effort to be reduced, yet at the cost of less planning accuracy. With demand forecasts being less reliable and less accurate over a long horizon, planning accuracy is less of an issue. A compromise is therefore possible between more accurate planning in smaller time buckets over the short horizon and less accurate planning in larger time buckets over the long horizon, with most of the planning effort being paid to the short horizon.

Planning horizon segmentation is a specific problem if not all time buckets are the same across the planning horizon. A possible planning horizon segmentation could for instance consist of a minimum of four weekly time buckets, followed by a minimum of three time buckets each of 4 weeks in length, followed by time buckets of 12 weeks in length. Figure 4.9 illustrates how the planning horizon segmentation changes when the planning horizon rolls through time, week by week. After 1 week, nothing fundamental changes. However, after one supplementary week, the first time bucket of 4 weeks is decomposed to four weekly time buckets in order to have a minimum of four weekly time buckets. The first time bucket of 12 weeks in length is decomposed to three buckets of 4 weeks in length in order to have a minimum of three such time buckets. Decomposition of time buckets involves decomposition of forecasts and MPS numbers.

W	W	W	W	W	4 weeks	4 weeks	4 weeks	12 weeks	12 W.

W	W	W	W	4 weeks	4 weeks	4 weeks	12 weeks	12 W.

W	W	W	W	W	W	W	4 weeks	4 weeks	12 weeks	12 W.

planning horizon →

Figure 4.9 Planning horizon segmentation as time progresses.

With terms having been discussed, a start can be made with a more extensive discussion of master production scheduling. The following section focuses on the relationships between the type and structure of the production system on the one hand and the shaping activities with respect to the MPS function on the other.

4.3.4 STRUCTURING FOR SUCCESSFUL PLANNING IN DIFFERENT PRODUCTION ENVIRONMENTS

4.3.4.1 Introduction

The existence of standard MRP logic as well as of techniques for execution scheduling and control might give the impression that a standard approach for manufacturing planning and control is possible, whatever the shape of the production environment. This way of thinking simply focuses on the production system itself and neglects the degree of interaction between the production system and its market.

Master planning is at the interface between the production system and the market and is entirely exposed to the specific characteristics of this interface. The way master planning is executed should ideally mirror the **desired** degree of interaction between the production system and the market. It is therefore by means of master planning that the production planning and control architecture should be customized to the specific needs of the company.

4.3.4.2 The decoupling point

In Chapter 1, companies were categorized according to their response to the market. Distinction was made between MTS, ATO and MTO companies. MTS, ATO and MTO companies primarily address different types of markets. The markets of commodity products require instant availability of products. These markets are addressed by MTS companies, which produce to stock, based upon forecasts, and try to limit risk by limiting the product range. MTO companies are prepared to provide very customized products, but start to produce only after receipt of a firm customer order. ATO companies position themselves in between and address primarily the markets of durable products. With an ATO approach, customer order lead times are minimized by dividing the value chain into two stages, i.e. a stage of module manufacture based upon forecasts followed by a stage of final assembly of customized products. Risk is minimized by modularizing products and standardizing modules as much as possible.

MTS, MTO and ATO companies essentially differ by a different position of the decoupling point. In Figure 4.10, a decoupling point is defined as a physical point in the value chain of the production system, which separates the investment stage from the realization stage.

Within the investment stage, operations are executed in response to firm and anticipated demand. Within the realization stage, on the other hand, production is against firm customer orders. The decoupling point determines the minimum customer order lead time. The actual customer order lead time will always exceed the lead time for executing the realization operations, since those are only executed after arrival of the customer order.

Distinction between the investment and realization stages is most clear with ATO companies. The stage of module manufacture is planned based upon forecasts (although some customer orders may already have been confirmed by customers). Modules are manufactured in anticipation of actual demands. Operations within the module manufacture stage are therefore investment operations and are situated before the decoupling point. On the other hand, final assembly operations are executed to satisfy firm demands. They are located after the decoupling point and are called realization operations. The decoupling point separates the module manufacture stage from the final assembly stage and corresponds physically with a stock of finished modules. A decoupling point **must** correspond with stocks somewhere in the value chain, as the simple result of the fact that forecasts – by definition – never are perfect. Whatever the quality of the prepared forecasts, forecasts will never be entirely substantiated by firm customer orders. The released orders for investment operations will result in component and subassembly stocks that will never entirely match with what is required to satisfy the firm customer orders.

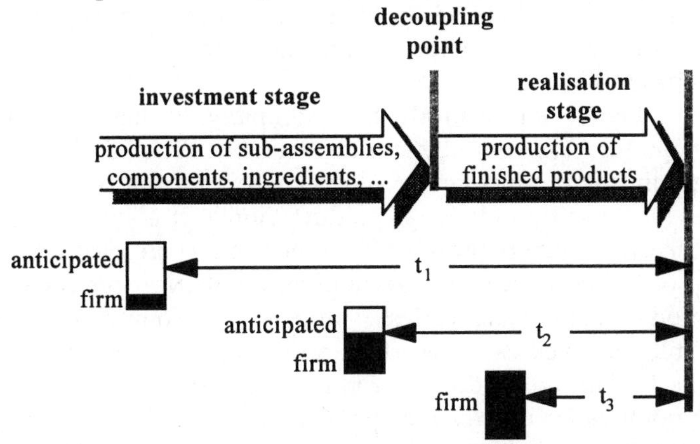

Figure 4.10 Typical mix of demand at time t_1, t_2 and t_3 from (expected) delivery.

Like ATO companies, which have their decoupling point between modular manufacture and final assembly, other companies also have their decoupling point at a specific point in their value chain. MTS companies produce to stock and may also prefer to execute part or all of the distribution operations based on forecast. The decoupling point is then accordingly positioned just after production, within or at the end of the distribution chain. MTO companies may have no decoupling point, with all activities driven by actual customer orders. However, to maximize responsiveness, they may prefer to set a decoupling point for long lead time items, which are then procured to forecast.

The position of the decoupling point determines the type of response of the production company to its market. It determines the level of interaction between production and the customer. The further upstream the decoupling point is positioned, the more interaction there is, since all of the operations positioned after the decoupling point are directly driven by the schedule of firm customer orders.

The position of the decoupling point should result from the guiding policies from business planning or might constitute a business planning decision itself. Its position determines the minimum customer order lead time that the company will provide to its customers, as well as the risk factor that the company is exposed to. The further downstream a decoupling point is positioned, the faster the company will deliver to its customers, but the higher will be the risk of obsolescence of materials that are produced on the basis of forecasts. The positioning of a decoupling point is therefore an important trade-off decision and should be made in the light of market uncertainty, service level expectations and the nature of the production process.

It is quite logical for a production company to have different decoupling points, e.g. one by product family. The choice of a decoupling point by product family allows the production company to optimize its response by individual product family. It is interesting in this respect to draw a comparison between the product life cycle and the progression of the position of the decoupling point. New products are typically produced in small series. There is low competition. The product is still perceived as a newcomer and long customer lead times are tolerated. It is in the production company's interest to position the decoupling point as far upstream as possible in order to minimize risk and to adapt its production volumes to match real requirements. However, as products become mature, competition steadily increases, demand increases, market shares become more visible and forecasts become more reliable. The company is actually forced to move its decoupling point downstream, in particular to keep its competitive position in terms of response and customer order lead times. The evolution continues in either of two possible directions. The company may prefer to go for a cost strategy. It tries to reduce cost by going for MTS mass production with a limited product range. The decoupling point moves further downstream. The alternative is to keep its competitive position by choosing a differentiation strategy. Product ranges are extended, while the required modules to manufacture the product variants are standardized as much as possible. The decoupling point is stabilized in between the module manufacture stage and the final assembly stage.

4.3.4.3 Optimal level of control

The presence of a decoupling point in a production system actually calls for two different schedules, namely the MPS and the realization schedule.

The realization schedule is a schedule of milestones to synchronize all operations within the realization stage, possibly via MRP. The realization schedule responds to firm customer orders, and is expressed at the level of finished products. It may deviate from the schedule of promised customer orders as a result of desired lot sizing or load levelling.

The MPS is a schedule of milestones driving MRP to plan and control all operations within the investment stage. The MPS responds to a mix of firm and anticipated demands. It is expressed in terms of those items produced by the investment stage.

In summary, investment operations are under the control of the MPS, realization operations are under control of the realization schedule. This is illustrated in Figure 4.11.

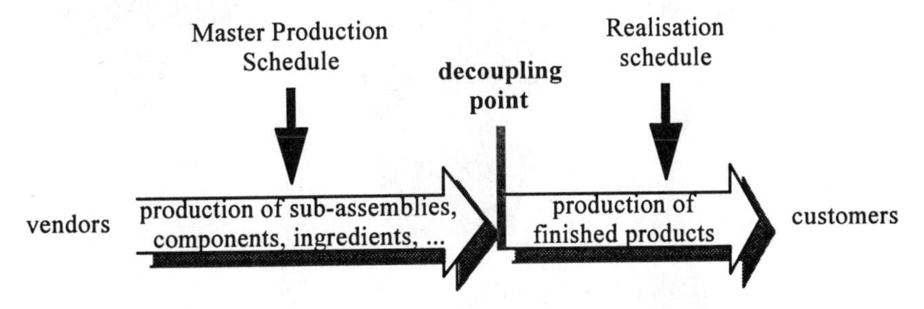

Figure 4.11 Role of MPS versus that of the realization schedule.

The realization schedule aims at satisfying firm customer orders. There is no uncertainty of demand. An order becomes only part of the realization schedule after acceptance of the customer order. In contrast, the MPS aims at anticipating demands. It is therefore at the level of the MPS that demand uncertainty needs to be managed. Management of demand uncertainty implies that safety inventories may be provisioned to be able to respond to unexpected demands. This idea will be further explored in the following section.

The MPS and realization schedule are clearly different in the case of ATO companies. The realization schedule is expressed at the level of finished products and corresponds to what is commonly named the final assembly schedule (FAS). The term FAS is only used within the ATO environment (the term realization schedule within this book is used as a more generic one). The FAS is a schedule indicating quantities and timing of products to be assembled starting from the modules made available by the MPS.

A business operating MTS has its decoupling point after all of the production operations. An MPS manages the responsiveness at the level of finished products. No realization schedule is involved. Note that in the MTS environment any (final) assembly operations are under control of the MPS, since they are executed in response to a combination of firm and anticipated demand.

All of the production operations for the MTO company are realization operations and are therefore driven by the realization schedule. Long lead time items may however be purchased based upon forecasts. It is therefore at the level of these purchase parts that responsiveness is managed. Responsiveness is typically managed by individual component, and no real MPS exists.

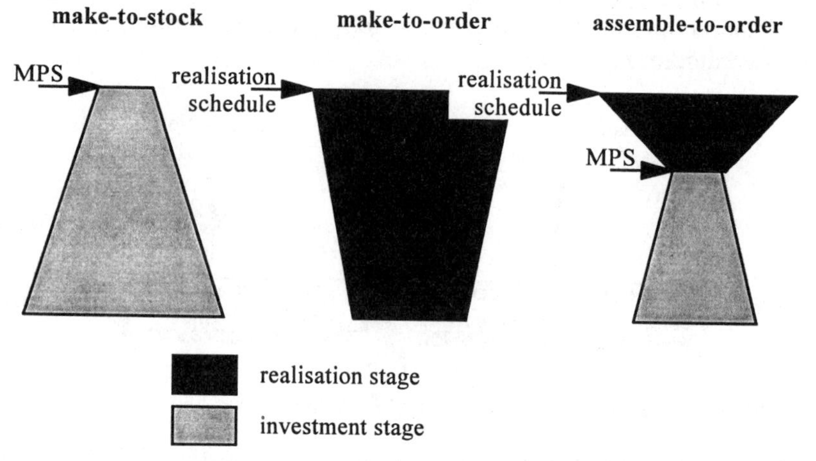

Figure 4.12 Ideal levels for realization schedule and MPS according to production approach.

Figure 4.12 identifies the desired levels for expressing the realization schedule and MPS, according to the production approach. The diagrams represent an idea of the item range to be dealt with from start of production (bottom of the diagrams) to end of production (top of diagrams). MTS companies typically produce a limited range of products starting from a more diverse set of components and raw materials. MTO companies provide the widest product choice. ATO companies produce a wide range of products starting from a standardized set of modules. The MPS is expressed in terms of those items produced by the investment stage. The figures make clear that this level corresponds with the level of highest commonality within the product structures (the level with the least number of different items). This again illustrates that responsiveness is best managed at the level of the MPS items. Buffering can then be limited to a minimum number of items, which maximizes the flexibility of the buffers and minimizes the cost of providing a certain degree of responsiveness.

It is worthwhile in this respect considering the situation of the MTS environment with the decoupling point halfway through the distribution network. Production operations and distribution to local warehouses are executed to anticipate demand. Distribution from local warehouses to final outlets is driven by firm customer orders. The choice of the decoupling point implies a policy of holding safety stocks at local warehouses. The policy is usually dictated by the market which demands short delivery lead times, e.g. shorter than those corresponding with distribution from production sites.

In practice, such companies will try to reduce the cost of the buffering policy, by keeping minimum safety stocks at local warehouses and some extra safety stock at the central finished goods warehouse. This is entirely in line with the advantage associated with buffering at the highest level of commonality. The safety stock at the central warehouse is much more flexible in its use than safety stocks at local warehouses, since no commitment to any warehouse has been made. Keeping safety stocks at local and central warehouses is a compromise between the wish to be highly responsive and at the same time cost-effective.

4.3.4.4 Relationships between final assembly scheduling and master production scheduling within the ATO environment

The MPS and FAS drive different sets of operations and are based upon different demand inputs. The FAS responds to firm demands over the final assembly horizon, i.e. the horizon covering the lead time of final assembly operations. The MPS responds to the combination of firm and anticipated demands beyond the final assembly horizon. Quite obviously therefore, the MPS and FAS can be developed highly independently from each other.

Yet, some relationships must be taken into account. The FAS can only respond to firm demands to the extent that product modules are available to assemble the products. Product modules must be checked upon availability before these can be reserved and issued. But availability of product modules should in the first place be guaranteed by accurate order promising based upon the MPS. Order promising will be addressed in section 4.3.6.

A second relationship is that the FAS generates module requirements for the MPS. The MPS should therefore respond to two demand inputs: firstly the statement of net demands summarizing the firm and anticipated demands beyond the final assembly horizon and secondly the MPS item requirements that derive from the FAS that are not yet satisfied. This is illustrated conceptually in Figure 4.13, and with a concrete example in Figure 4.14.

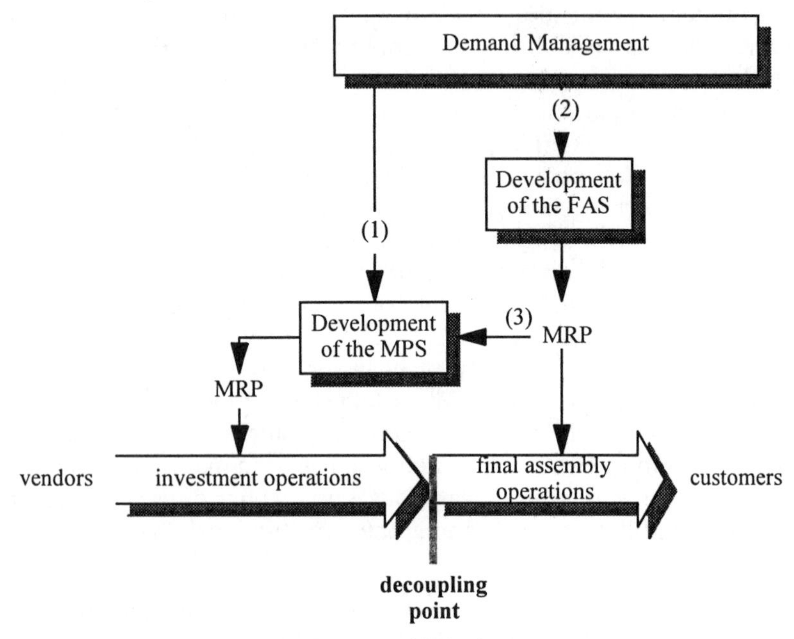

Figure 4.13 Inputs to the development of the MPS.

The concrete example is for a product X that is assembled from – among others – one module Y. The product X is under control of the FAS. The module Y is an MPS item. For purposes of simplicity it is assumed that there are no intermediate levels between item X and item Y in the bill of material for product X. The example assumes a final assembly lead time of 2 weeks. FAS and MPS orders are scheduled to be completed at the start of the relevant time buckets. For example, the net demand for week 3 should be covered by an order due to arrive at the start of week 3, which – when taking account of the final assembly lead time of 2 weeks

Product X	p. due	week 1	week 2	week 3
Customer orders		10	10	10
FAS		8	10	10
Scheduled receipts		8	10	
Pl. order releases		10		
PAI	2	0	0	0

On hand = 2, final assembly lead time = 2 weeks

X

Y

Demand management for MPS item Y, before offset by final assembly lead time

MPS item Y	p. due	week 1	week 2	week 3	week 4	week 5	week 6
Forecast					10	10	10
Customer orders					8	4	
Net demand					10	10	10

Master Production Schedule for MPS item Y

MPS item Y	p. due	week 1	week 2	week 3	week 4	week 5	week 6
Total requirements		10	10	10	10		
MPS		20		20			
PAI	5	15	5	15	5	5	5

On hand = 5

Figure 4.14 Inputs to the development of the MPS (concrete example).

– is to be released at the start of week 1. A final assembly horizon of 3 weeks is therefore required.

The length of the final assembly horizon typically corresponds to some realistic estimate of the final assembly lead time (in this example 2 weeks). In the case of constrained capacity within final assembly, a longer final assembly horizon may be provided to allow the final assembly scheduler to consolidate production orders in order to minimize set-up.

The first table shows the final assembly schedule for product X over the final assembly horizon. One order of 10 units is yet to be released. This results in dependent requirements for MPS item Y (10 units in week 1). It is assumed that the components for all open orders have been issued.

The second table shows the calculation of the net demand for MPS item Y beyond the final assembly horizon. The net demand calculation is part of the demand management function that compares firm customer orders against total forecast. This is discussed in later sections of this chapter. It must be stressed that forecasting may not necessarily be done at MPS item level or at finished product level. In the ATO environment, it is much more practical to forecast demand and calculate net demand by product family, and to derive – by straightforward explosion – the corresponding forecasts for the individual MPS items.

Returning to the example, a net demand of 10 units in week 4 (table 2) means that 10 units of MPS item Y are expected 'to be sold' through various product sales during week 4. To enable the products to be assembled in time, the net demand for item Y must be offset by the final assembly lead time. Finally, total requirements derive from the dependent requirements of the final assembly schedule, and from the net demand statement beyond the final assembly horizon.

Unfortunately, many ATO companies still manage their business as if it were an MTS environment. Firm and anticipated demands are expressed for the individual finished products over the **entire** planning horizon, and a 'supply schedule' is developed for each of the finished products. Both the realization and investment stages are controlled at the level of finished products. The supply schedules drive MRP. MRP calculates the corresponding dependent requirements for the individual product modules.

This approach is characterized by two serious shortcomings. Firstly, forecasting is not optimized since it is done for the individual products and not for product families as is much more realistic. Secondly, there is no satisfactory basis for order promising. Order promising should ideally be executed at the level of the product modules. On the one hand, this level is the level of lowest commonality and therefore the ideal level for 'provisioning' safety. On the other, incoming customer orders should be promised against (projected) available product modules, since those are the items produced to forecast, anticipating actual demand. In contrast, with product schedules being developed at the level of finished products, order promising is actually executed at that same level. Each incoming customer order is verified against the corresponding supply schedule. ATO companies have sometimes thousands of finished product variants, which can mostly be categorized in product groups of similar product variants. Supply schedules are therefore only developed for the more commonly sold finished products. If a less commonly sold finished product is ordered, a corresponding supply schedule may not exist. As a consequence it gets promised against the supply schedule of a related finished product, assuming that those materials that are different from the chosen finished product have been planned for in the past and will be available to support the production of the accepted order. Also, because of the high number of different finished products, only a limited amount of overplanning can be accepted for each of them. The result is that with slight deviations between actual demand and forecasted demand the in-built reserves do not appear to be sufficient and incoming orders have again to be promised against the supply schedules of related products. The same remark holds true.

4.3.4.5 Planning bills of material

The above text showed that the planning approach should be consistent with the type of the production environment. MPS items should correspond with those items produced by the investment stage. These items are typically situated at the level of lowest commonality within the product structures, often corresponding with a suitable decoupling point in the production process. The text also indicated that, in contrast with the realization schedule, the MPS is developed in response to the combination of actual and anticipated demand beyond the horizon of the realization schedule. Although forecasts are required for each of the MPS items, it is often much more practical to develop the basic forecasts at a higher, more aggregate level (e.g. for product families).

These two elements, the possible choice of MPS items at an intermediate level within the bills of material and the choice of a forecasting level at a higher, more aggregate level, often lead to the need for the use of planning BoMs.

Modular planning bills of material

A planning bill of material can be defined as any bill of material that is used to facilitate planning and control as opposed to bills of material used for producing a product. Planning bills of material are extremely relevant in ATO environments where the bills of material are restructured in such a way that only a limited number of modules have to be planned as part of the master production scheduling process. The identified modules are not necessarily buildable. They are defined to facilitate master production scheduling.

Consider the production of personal computers (PCs) as an example. A certain company produces PCs that come in two hardware configurations (tower or compact), two versions of CPU power (386 series and 486 series), six screen types (black and white and colour with low, medium or high resolution), three possible configurations for the internal memory, and two configurations for the hard disk memory. This results in 144 ($= 2 \times 2 \times 6 \times 3 \times 2$) possible combinations. Rather than building each of the PCs to stock, the company prefers to build or purchase the standard modules to forecast (such as the frame for the tower version, the frame for the compact version, the different screen types, etc.) and to assemble them to a specially configured PC system upon entry of a customer order. The possible combinations are visualized by the superbill structure of Figure 4.15.

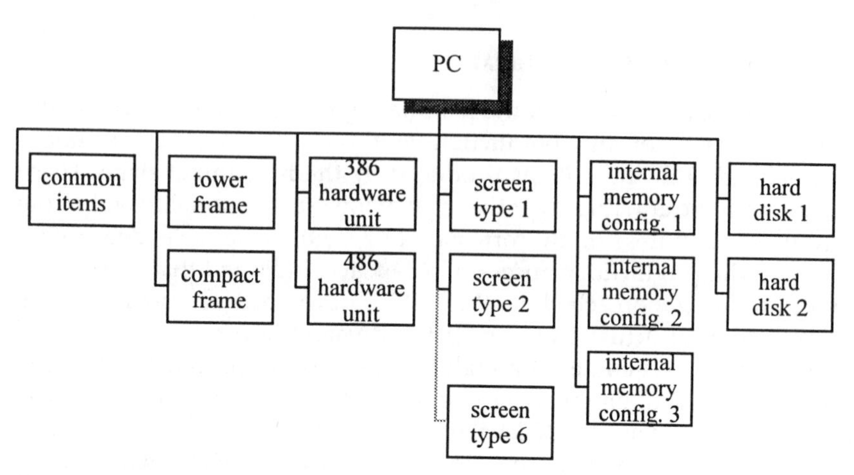

Figure 4.15 Superbill structure for a PC.

Master production scheduling is done for the 16 identified modules. Note that not all of the modules are buildable. The common parts set for instance may only represent a kit of parts that are required whatever the special configuration of the PC. Nor are the different screen-type modules buildable modules. They may consist of the screens themselves as well as – among others – the corresponding screen cards that have to be assembled within the PC frame. Final assembly of the PC should therefore be dictated by a manufacturing bill of material or M-BoM, which identifies – by PC variant – how to assemble the computer.

The method of defining the modules and structuring the bills of material should ensure the number of MPS items is minimized, and the MPS is expressed in terms of items that have a direct relationship with the way the product is sold. This enables a much easier integration between master production scheduling and demand management. Product family forecasts and customer orders can easily be translated (e.g. again using planning bills of material) to corresponding MPS item forecasts/requirements. Customer order promising is facilitated (against supply schedules for 16 modules rather than for 144 finished products) and it becomes easier to plan safety inventories.

Superbill structures

Another form of planning bill is the superbill. Superbills intend to limit the number of items to master schedule by grouping a number of modules under one (non-engineering) part number. Figure 4.16 illustrates the creation of a superbill structure for a generic screen that by means of different percentages is defined as a mix of different screen types. Master production scheduling is done for the generic screen rather than for the individual screen types. The percentages are typically determined by analysing historical demand by individual screen type. The sum of percentages should equate to 100 in the case of each PC being sold with one screen. Otherwise, the sum may be less (or more) than 100.

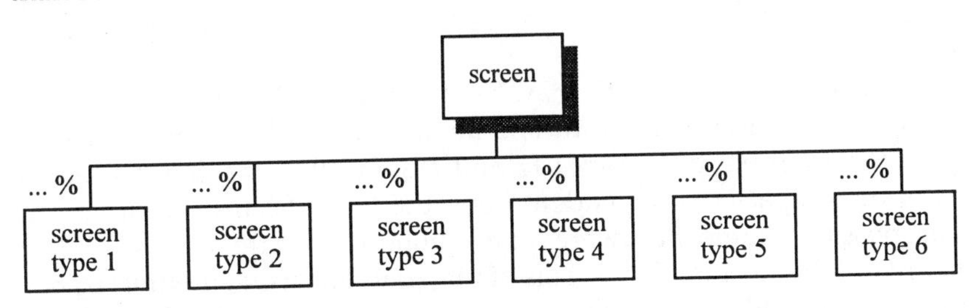

Figure 4.16 Superbill structure for a PC screen.

The use of superbill structures would only be useful if the percentages – determined by historical analysis of demand – remain quite stable over time. Further, some safety needs to be planned to account for short-term deviations from the expected percentage averages. It is not recommended to plan safety by overplanning using higher percentage figures (so that the sum of percentages would be higher than 100), but to plan safety using safety stock targets for each of the individual modules.

Modularizing bills of material

Modular bills of material reflect the modular design of a product family, which enables the production operations for the modules to be 'decoupled' from the assembly operations for the products. Some companies however evolve gradually from an MTS or MTO approach to an ATO approach and – at a certain moment in time – try to modularize bills of material and adjust product design in order to be able to take advantage of the ATO approach.

The process of modularizing consists in breaking down the bills of products and restructuring them into separate bills for the individual modules. In the case of a good modular design, the number of modules is much less than the total number of possible product combinations (144 possible product combinations versus 16 modules for the example of Figure 4.15). This results in major advantages. Firstly, it is much easier to develop forecasts for each of the modules than for each of the products. More accurate forecasts mean fewer revisions, necessitating fewer changes in the master production schedules. Secondly, safety quantities can be planned for the individual modules, rather than for the products. Safety quantities at module level are not yet committed to a certain variant of the product family, and help to address demand uncertainty properly. A good modular design permits the commitment of materials (subassemblies, modules) to specific variants of the product family to be postponed until as late as possible in the production process.

Product 'modularization' may be complicated by the existence of items that are used only with specific option combinations. Consider again the example of Figure 4.15 and assume that the frame – either tower or compact – is dependent upon the type of hardware unit. The four possible frame–hardware unit combinations are represented in Figure 4.17 by means of their bills of material. The usual modularization approach would try to allocate the different constituent items to individual modules. This allocation is illustrated in Figure 4.18 for the tower frame–386 hardware unit combination. Items A, B and D can be allocated to individual modules. Item F however is specific to the tower frame–386 hardware unit combination and cannot be allocated to an individual module.

There are three possible approaches.

- Master production scheduling is done for the four possible combinations of frame units and hardware units, and not for the individual frame and hardware unit modules. The disadvantage is that combinations of modules are more difficult to forecast, leading to more forecast revisions and revisions of the master production schedules. In addition, buffer inventories planned for the combinations of modules are dedicated to those combinations and cannot be used for other combinations.
- Master production scheduling is done for the four individual modules, and not for the combinations of modules. This, however, requires that the items that cannot be allocated to a module be planned separately.
- The items that cannot be allocated to a module are further exploded to their constituent items, with the intention of mapping these to the individual modules. This is only possible if the items are subassemblies of other items, and if the corresponding subassembly operations can be scheduled as part of the final assembly scheduling process. If the items have to be assembled before final assembly

(mainly because of customer lead time considerations), then there is no point of diving deeper in product structures to find a higher degree of modularity. The combinations of options have to be produced in anticipation of actual demands and therefore have to be master scheduled as such.

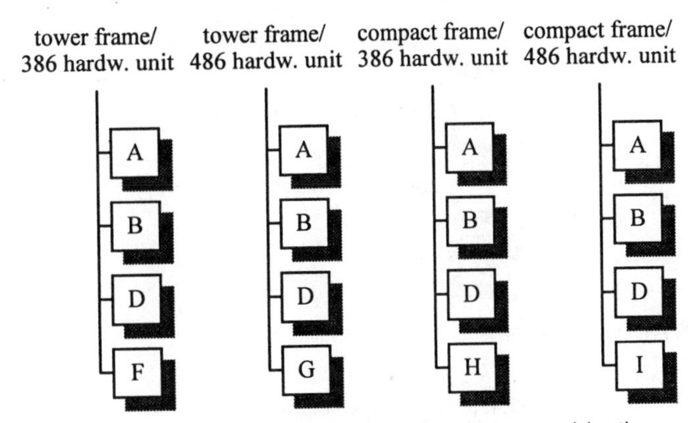

Figure 4.17 Bills of material for the four frame/hardware combinations.

Example case: a manufacturer of digital exchanges

This text describes the case of a major manufacturer of digital phone exchanges. Digital telephone exchanges are highly modular products. Figure 4.19 represents a simplified structure of the bills of material. The company is a pure ATO environment. Pre-equipped racks, power supplies and subassemblies are produced to forecast, whereas the final assembly of printed board assemblies (PBAs, more often referred to as printed circuit boards or PCBs), racks and digital exchanges is driven by firm and fully configured customer orders.

The power supplies, pre-equipped racks and printed board assemblies are the basic building blocks of the digital telephone exchange and have quite intuitively been chosen as the MPS items. As mentioned before, the power supplies and pre-equipped racks are produced to forecast. This involves limited risk, since both power supplies and pre-equipped racks are quite standard modules that can be used across a wide range of digital exchanges. As regards PBAs, there exist over 200 variants with a high degree of commonality. Specific digital exchanges require a specific combination of different PBAs. In order to maximize responsiveness to changes in demand (primarily configuration changes), manufacture of PBAs is postponed until the final assembly stage. So, although the master production schedules are expressed in terms of the PBAs, release of the planned orders is under control of the FAS.

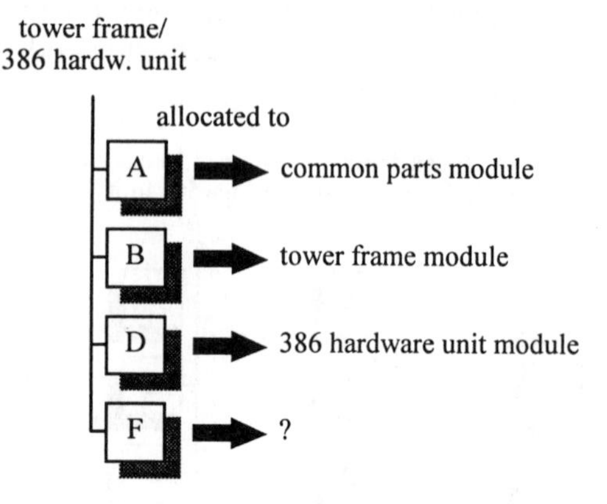

Figure 4.18 Allocation of items to individual modules.

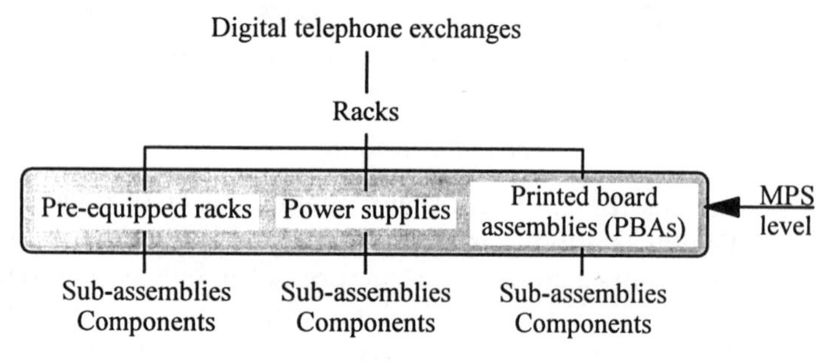

Figure 4.19 Simplified structure of bills of material of digital exchanges.

The MPS exercise is heavily focused on managing demand uncertainty and planning buffer inventories. Buffer inventories are required to optimize the response of the production system, particularly in light of the long purchase times for components of PBAs.

The considered company has started an interesting exercise, aiming at minimizing investment in buffer inventories by reconsidering the items for master production scheduling. In the current production approach, components and subassemblies required for assembly of PBAs are only committed during the final assembly stage. The release of planned orders for PBAs is indeed under control of the FAS. This constitutes an element of flexibility that is not reflected by the current master production scheduling approach. Each PBA is master scheduled individually, and buffer inventories are planned by PBA item. The commonality among the different PBAs is not visible at master production scheduling level. PBAs are master scheduled as if components and subassemblies were committed to the individual PBAs. The feasibility of decreasing certain master production schedules in favour of increasing master production schedules of like PBAs can only be assessed by MRP simulations.

It is clear from the above that current master production scheduling is not executed at the optimum level, i.e. the level of highest commonality. The set of PBAs can actually be categorized in a series of PBA families. PBAs within these PBA families are characterized by a high degree of commonality in terms of constituent subassemblies or components, which can be grouped into common and specific item sets. Common item sets are kits of subassemblies or components that are common to a certain family of PBAs. Specific item sets are kits of subassemblies or components that are specific to a particular PBA. Subassemblies are not exploded to determine commonality at a lower bill of material level, since customer lead times require that these be manufactured in anticipation of the realization schedule and therefore be driven by the MPS.

The objective of the commonality analysis was to arrive at bills of material as represented in Figure 4.20, with master production scheduling executed at the level of pre-equipped racks, power supplies and common and specific item sets. This planning approach reflects the current production approach. Buffer inventories can be planned at the level of common and specific item sets, and are not yet allocated to a specific PBA variant. The allocation of common and specific item sets to PBAs can be postponed until final assembly. Not only can the cost of buffer inventories be reduced to provide a certain degree of responsiveness (buffer inventories are applicable for different PBAs), but also the commonality among PBAs is visible at the master production scheduling level. No further MRP simulations are required to check the feasibility of a swap of MPS quantities between like PBAs.

There was no straightforward grouping of PBAs. Different grouping scenarios were compared with each other to minimize the cost of buffer inventories. The cost of buffer inventories was estimated by taking into account a predetermined responsiveness policy. This responsiveness policy identified targets for PAI over the planning horizon, dependent upon the sales expectations of each individual MPS item. The higher the sales expectations, the lower the targets for PAI (expressed as a percentage of average sales expectations). The rules reflect the idea that the higher the sales expectations, the higher the forecasting accuracy, and the lower the need for buffering is expected to be (proportionally with respect to the sales expectations).

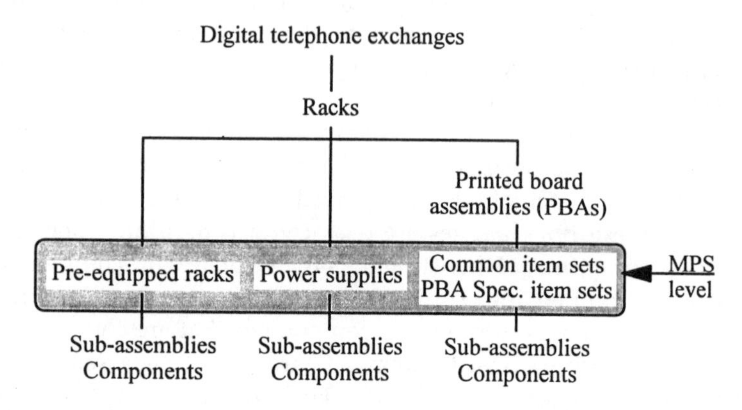

Figure 4.20 Proposed structure of bills of material of digital exchanges.

The responsiveness policy was first applied on the PBA items themselves. The sales expectations for each individual PBA item result in targets for projected available inventory. These targets for projected available inventory at MPS item level result in overplanning within the cumulative procurement and production lead time, which in turn results in safety stocks being carried at lower levels of the bills of material. The safety stocks have a cost, and this cost can be estimated by analysing the bills of material of the MPS items.

The same responsiveness policy was then applied to each considered grouping alternative. Each alternative corresponds with a set of specific and common item sets, depending upon the PBA families chosen. Buffering is no longer done by MPS item, but can be done by specific and common item set. Sales expectations for common item sets are higher than those of the individual PBAs. With higher sales expectations, buffer levels are set at a lower level (again proportionally with respect to the sales expectations) and cost of buffering can be reduced. Total investment in buffer inventories is therefore lower when planned at the level of common and specific item sets than when planned at the level of PBAs. The definition of a responsiveness policy and the corresponding calculation of cost of buffer inventories allowed the comparison of different grouping scenarios and the selection of the one with the minimum associated cost.

The change from PBAs towards common and specific item sets corresponds with a more logical choice of the position of the MPS items within the bills of material, as represented in Figure 4.21.

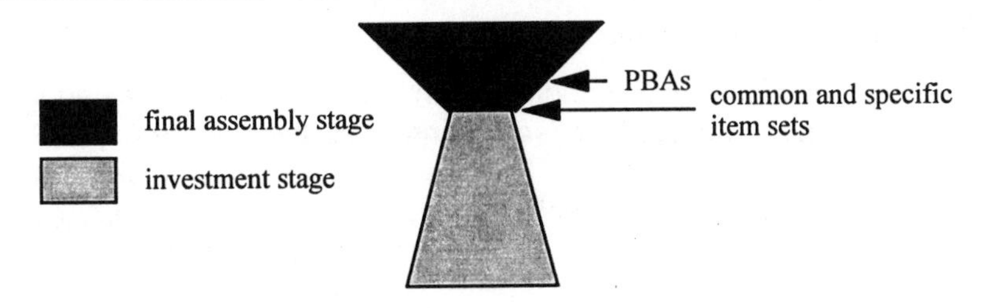

Figure 4.21 Position of PBAs and common and specific item sets within structure of bills of materials of digital exchanges.

4.3.5 DRIVING THE PRODUCTION SYSTEM THROUGH MASTER PLANNING

In this section, the MPS will be discussed from the perspective of its role as a driver of MRP within the MRP II-based planning and control architecture. Its second role, i.e. as the basis for order promising, will be discussed in section 4.3.6.

The MPS plays an important role as a buffer between the production system and its limitations in terms of responsiveness on the one hand and the variability of demand on the other. The MPS is therefore the prime planning vehicle for managing the responsiveness of the production system to its market. This idea will be explored in a first part of the text.

The subsequent parts focus on the development of the MPS. A feasible MPS implies that it be developed within the boundaries of material and capacity availability. The discussion focuses on the commonly accepted approach within the standard MRP II framework.

4.3.5.1 Management of responsiveness

The MPS is developed in response to a combination of firm and anticipated demands. This demand profile may change, and ideally the MPS should be changed accordingly.

Decreasing or de-expediting MPS order quantities is possible at any time. The result is likely to be a temporary increase in stock. The temporary increase in stock results from open and completed production and purchase orders for subassemblies and components, which had been planned by MRP to satisfy the originally planned MPS item orders. However, increasing or expediting MPS order quantities is only possible to a limited degree, determined by the possibilities to reduce lead times, either externally (with suppliers or subcontractors) or internally (within production), or by the incidental availability of sufficient inventory at lower levels of the product structures.

As Orlicky (1974) has already indicated:

> . . . the limits of flexibility contract with passage of time, making it less and less practical to effect changes as the end item (MPS item) nears its scheduled completion date. The reality of commitment acts as a funnel with ever-narrowing walls that, as time goes on, leave less and less room for deviation from original plan.

Buffer inventories must therefore be planned to provide a desired degree of responsiveness. The MPS is the preferred vehicle to protect the production system from demand uncertainty. If inventories are used to increase responsiveness, they should preferably be planned and be visible at the level of the MPS. Planning inventories at the level of the MPS ensures that safety investments are consistent across all components and subassemblies. Additionally, overplanning is visible as part of the MPS, which is a necessary condition for efficient order promising.

Items under control of MRP are managed as dependent upon their parent items. The demands can be derived from the planned replenishments at higher levels of the product structures. Safety stocks for such items should be minimized as much as possible, since requirements are known. Those requirements should remain quite stable over time, in light of the recommendation to buffer the production system from unexpected changes in demand at the level of the MPS items. Safety stocks, if any, for MRP controlled items should primarily be used to protect the production system from uncertainties with respect to supply (from external suppliers or from preceding production steps).

An interesting concept to control and ensure responsiveness at the level of the MPS items is the concept of **hedging**. Hedging is a technique to provision inventory buffers for the individual MPS items at certain time fences within the planning horizon. Rather than planning safety stock for each of the MPS items (which may prove to be expensive), hedging aims at planning 'time-phased' safety stock. The planned buffers at the predefined time fences serve to guarantee a certain degree of responsiveness with a certain delay as determined by the time fences at which they are planned. In case the buffers are not required, they are de-expedited with every planning cycle, so that they are positioned again at the same 'time distance'. If buffers are required, planners allow these buffers to approach with every planning cycle. The MRP explosion process ensures that all required components and subassemblies are kept in stock to guarantee an extra supply of the concerned MPS item within a time delay as determined by the time fence at which the buffer is planned. A numerical example can be found within the insertion.

De-expedition of hedges at the MPS item level results in de-expedition of planned orders at all of the lower levels of the product structures. De-expedition of non-released planned orders for purchase parts could be considered a problem when using the output of MRP as the basis for supplier schedules. The output of MRP does include planned quantities corresponding with the hedges planned at the MPS item level. Use of hedges in combination with supplier schedules should be managed by supplier agreements defining the time fences within which planned orders should be considered firm orders. These time fences indicate when planned order releases are called off. As such, they determine the purchase lead times within the overall supplier agreements, as well as the degree to which overplanning results in safety stocks carried with the production company.

The concept of hedging is entirely in line with the need for providing more safety margin in the distant planning horizon as opposed to the nearby horizon. The time-phased buffers result in higher PAI figures in the distant horizon. Two good reasons for having a step PAI profile may be indicated. One reason is that demands are more difficult to forecast the more distant they are. The MPS should therefore provide more 'safety margin' in the distant future in order to be able to accommodate a higher degree of demand variations. A second reason is purely economic in nature. Hedges are more expensive the more nearby they are positioned. Hedges result through MRP explosion and after successive weeks of planning in safety stocks for dependent subassemblies and components. As hedges are positioned more nearby, safety stocks will be carried at higher levels of the product structures and will prove to be more expensive. There is therefore an advantage in positioning hedges only as nearby as necessary to provide the required responsiveness.

The above is in sharp contrast with the tendency of many companies to adjust the in-built safety margins of their master production schedules (as reflected by the projected available inventory figures) towards zero towards the end of the planning horizons. This is only justified to the extent that the considered MPS item is at the end of its life cycle, and safety margins have to be scarcely planned.

Insertion: Hedging: an example

The concept of hedging is illustrated in Figure 4.22. For the purposes of the example, the MPS is divided into two parts, i.e. a first part that intends to meet firm and anticipated demand and a second part that raises the projected available inventory profile at specific time fences (2 units in week 3 and 2 units in week 5). The MPS numbers of the second part are called hedges. They correspond with 'safety quantities' not immediately available (as opposed to the on-hand stock figure of 2 units), but positioned a certain number of weeks from completion. When the demand profile remains stable, those safety quantities are rolled forward, week by week, and positioned at the same 'time distance'. The hedges are de-expedited. However, if the demand profile changes, the provisioned safety can be called upon to respond quickly to satisfy changed or new demands. The inventory buffers that are positioned at certain time fences result via explosion by MRP in matched safety quantities for the different subassemblies and components of the MPS items. Because the safety quantities are not held as safety stocks for finished products, safety quantities are less expensive to keep. The size of the inventory buffers and the position of the time fences can be determined in an optimum way so to achieve a good balance between responsiveness and cost of responsiveness. Ideally, these parameters should be determined in line with a pre-established responsiveness policy.

MPS item X	1	2	3	4	5
Net demand	10	10	8	8	10
MPS-1	10	10	8	8	10
MPS-2 (hedges)			2		2
PAI	2	2	4	4	6

On hand = 2

Figure 4.22 Using hedges to satisfy a responsiveness policy.

Figures 4.23 and 4.24 illustrate how the positioning of the hedges determines the safety stocks at lower levels of the product structures. One-for-one bill of material relationships are assumed. Figure 4.24 is the update of Figure 4.23, 1 week later. Since net demand has not changed, hedges are de-expedited from weeks 3 and 5 to, respectively, weeks 4 and 6. The de-expedition ensures that the projected available inventory profile moves along the time scale. It actually implies a change to the total MPS (MPS-1 + MPS-2) which results in changes of planned order releases. The planned order release of week 2 has reduced from 10 units to 8 units, whereas the one for week 3 has increased from 8 units to 10 units. Similar changes are true for the planned order releases in weeks 4 and 5. Gross requirements for item Y change as a result. The gross requirement of 8 units in week 2 combined with a scheduled receipt of 10 units will result in an expected stock receipt of 2 units by the end of week 2. For item Z, 2 units are expected to be received in stock by the end of week 3.

MPS item X	1	2	3	4	5
Net demand	10	10	8	8	10
MPS-1	10	10	8	8	10
MPS-2 (hedges)			2		2
PAI	2	2	4	4	6
Scheduled receipts	10				
Pl. order releases	10	10	8	12	

On hand = 2, lead time = 1 week

item Y	1	2	3	4	5
Gross requirements	10	10	8	12	
Scheduled receipts	10				
PAI	0	0	0	0	
Pl. order releases	10	8	12		

On hand = 0, lead time = 1 week

item Z	1	2	3	4	
Gross requirements	10	8	12		
Scheduled receipts	10	8			
PAI	0	0	0		
Pl. order releases	12				

On hand = 0, lead time = 2 weeks

X
|
Y
|
Z

Figure 4.23 Impact of hedging for MPS item X at lower levels.

The hedges for the MPS item X are rolled forward, week by week. As a result, a stock of 2 units will continue to be carried for item Y, as well as for item Z. These stocks determine the responsiveness of the manufacturing system to changes in demand. In case the net demand for the MPS item X would increase, the hedges would not be rolled forward but used to meet extra demand. Extra orders should then be planned beyond the cumulative lead time to adjust actual PAI figures with target PAI figures. A decreasing net demand would result in de-expedition of more than simply the hedges. Existing open orders (scheduled receipts) may cause the intermediate stocks to go up temporarily before they are readjusted by the MRP logic.

Note that the hedges of 2 units positioned in the third and fifth week in the planning horizon result through MRP in safety stocks being carried for items Y and Z. MRP makes sure that the provisioned safety within the MPS is translated down consistently to lower levels of the product structure.

It is important to note that the responsiveness of the production system is determined by the intermediate stocks, and yet entirely driven by the responsiveness policy for the MPS item X. Responsiveness is entirely visible and creates the ideal environment for efficient order promising.

MPS item X	p. due	2	3	4	5	6
Net demand		10	8	8	10	10
MPS-1		10	8	8	10	10
MPS-2 (hedges)				2		2
PAI		2	2	4	4	6
Scheduled receipts		10				
Pl. order releases		8	10	10	12	

On hand = 2, lead time = 1 week

item Y	p. due	2	3	4	5	6
Gross requirements		8	10	10	12	
Scheduled receipts		10				
PAI		2	0	0		
Pl. order releases		8	10	12		

On hand = 0, lead time = 1 week

item Z	p. due	2	3	4	5	6
Gross requirements		8	10	12		
Scheduled receipts		8	12			
PAI		0	2	0		
Pl. order releases		10				

On hand = 0, lead time = 2 weeks

Figure 4.24 Planning table of Figure 4.23 after 1 week.

4.3.5.2 Material constraints

Material constraints are constraints in terms of material availability, and result from the purchase and production lead times. Since component item orders must be completed before the parent item orders (that will consume the component items) can be started, purchase and production lead times result in a cumulative lead time for each of the MPS items. Each individual MPS item has a specific cumulative lead time, which is defined as the lead time corresponding with the 'critical path' between start of procurement and completion of the MPS item order. It defines the minimum planning horizon for the MPS item. The minimum planning horizon may be identified by a time fence, which divides the planning horizon into two segments, a provisioning horizon and a tentative horizon. The time fence itself could be called the provisioning time fence. Each of the MPS quantities before the provisioning time fence is positioned within the cumulative lead time of the MPS item, and therefore corresponds with some committed-to investment, in the form of released purchase and/or production orders (Figure 4.25).

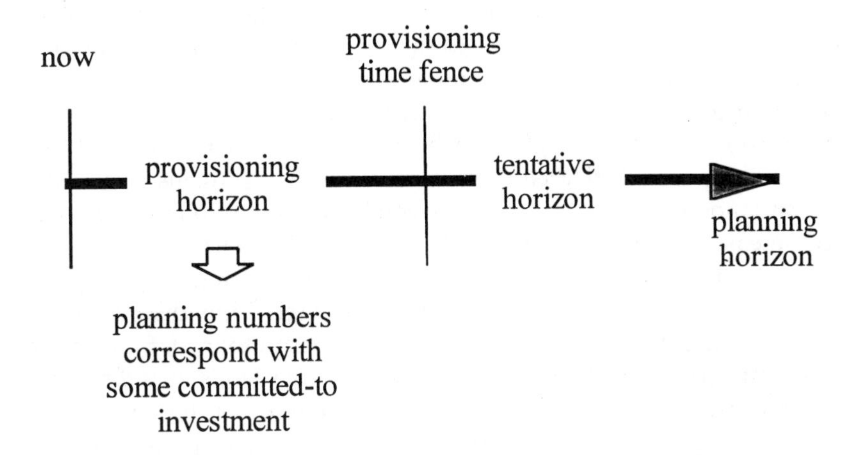

Figure 4.25 Planning horizon segments.

Ideally the MPS should be developed according to some buffering policy so that it can absorb a reasonable amount of unexpected demand variations within the provisioning horizon. Too often, however, the MPS needs to be changed within the provisioning horizon. When it comes to increasing the MPS quantities or expediting the MPS item orders, the master production scheduler must realize that not enough material may be in the pipeline (in terms of released purchasing and/or production orders) to support the proposed changes. Only an MRP simulation can show the feasibility of the proposed changes. Practice shows that shortages can quite often be resolved. Lead times in MRP tend to reflect worst-case situations and can often be reduced. Also, subcontractors or overtime can sometimes be called upon.

The master production scheduler may also consider decreasing the master production schedules for some MPS items in order to increase other master production schedules for like MPS items, within each of their provisioning horizons. In doing so, the master production scheduler simply hopes that the increased material requirements for the latter MPS items are compensated by decreased material requirements for the former ones. This again can only be checked by means of an MRP simulation run, which for many companies is time-consuming and often not considered when developing the master production schedules.

Note that appropriate structuring as discussed in the previous section may take advantage of the commonality among different products, and create the necessary conditions to develop and maintain the MPS item schedules individually. An example was given before with the case study on the digital exchange manufacturing company. Current master production scheduling is at the level of printed board assemblies (PBAs), which have a high degree of commonality among themselves. Often, the situation calls for decreasing the MPS quantities of one PBA in order to increase the MPS quantities for another PBA. By moving from the current set of MPS items to a set of MPS items of common and specific item sets, each MPS item could be considered individually without much overlap in terms of material content with other MPS items. The actual material availability in terms of PBA components would be much better reflected by the master production schedules for the common and specific item sets. The feasibility of meeting unexpected demand changes could be derived directly from the planning tables for the individual MPS items.

4.3.5.3 Capacity constraints

The MPS results in decisions for procurement and production, and as such should be developed within the boundaries of available productive capacity. In this respect, it is important that the capacity policy – as developed by the business planning activity – is taken into account. The capacity policy indicates management's commitment to invest in productive capacity and indicates the extent to which subcontracting can be resorted to.

Overstated master production schedules result in increasing component inventories and work in process, because more components are procured and more work is released to the shop floor than what can actually be feasibly processed. Queues grow and have an adverse impact on lead times. With lead times becoming unrealistic, the whole MRP-based production planning system may be jeopardized. The management of capacity has in this respect caused even more challenging problems to production management than the management of materials.

4.3.5.4 MPS development approach

The MPS has clearly to be developed within the boundaries of projected materials availability and available productive capacity. Ideally, both material and capacity constraints should be taken into account simultaneously during development of the MPS. This may be possible by considering the specific nature of the material and capacity constraints, and developing a corresponding customized development approach. However, a generic approach has not been developed yet.

Important is the fact that most companies are faced with constraints that are primarily either material or capacity constraints. These companies prefer to develop a tentative MPS while taking account of their primary constraints and subsequently verify the tentative MPS with respect to the secondary constraints. In this way, the company is informed about any overloads with respect to the latter constraints and may prefer to either change the tentative MPS or try to remove the secondary constraints, which are often less stringent in nature.

The MRP II-based planning architecture is primarily suited for companies faced with material constraints as the primary constraints. The logic of material requirements planning aims at planning and controlling material supply throughout the supply chain, while taking account of the dependencies between child and parent items as well as the purchase and production lead times. Not surprisingly, the capital goods industries and durables industries appeared to be the industries with the highest potential for applicability of the MRP II-based planning architecture. These industries are faced with complex bills of material and typically with long purchase and production lead times. They provide a wide product offering, usually MTO or ATO. Material supply is complex and constrains the response to the market. In order to optimize production lead times and not to be fully dependent upon a mix of material and capacity constraints, many of these companies strategically hold some degree of excess capacity. The primary capacity constraint is often within the final assembly stage of production.

It should therefore not be a surprise that the traditional approach for development of the MPS starts with the development of a tentative MPS within the boundaries of material availability, which subsequently is evaluated with respect to overall productive capacity. The approach can actually be summarized in three steps, which are:

- the identification of the demands placed upon the production system;
- the development of a tentative MPS, in accordance with the responsiveness policy (buffer inventories) and within the boundaries of material constraints;
- the rough-cut evaluation of the tentative MPS, from a capacity point of view.

The second and third steps may be repeated several times until a satisfactory MPS is arrived at. Any constraints within the final assembly/ realization stage are often taken into account as part of the second step of this process. This check is often easy to perform. It often relates to the total number of products that can be coped with within the realization stage. For example, for the digital exchange manufacturing company mentioned before, there is an upper limit on the total number of PBAs that can be assembled within the realization stage.

4.3.5.5 Identification of demands

The MPS represents the response of the production system to the demands placed upon the production system. These demands are not necessarily the external demands placed upon the company. In the MTS environment with a distribution organization, the demands are those of the distribution organization and not those placed by the end customer on the distribution organization.

The demands may typically include customer requirements, warehouse replenishment requirements, distribution requirements, interplant requirements, as well as the hedges to satisfy a predefined responsiveness policy. Any hedges required to implement a predefined responsiveness policy can be considered as demands. These demands do not originate from the production system's customer, but from business planning that aims at guaranteeing a sufficient degree of responsiveness.

Distribution requirements may be fed to the master production scheduling system by means of the distribution requirements planning (DRP) logic, which may be used to consolidate demands from different (warehouse) sources or from different business units. The DRP logic is actually used to create visibility as regards demands placed upon the production system starting from requirements at local warehouses or different business units (Martin 1990).

Requirements are either firm or anticipated. Firm demands are expressed as demands for finished products. Anticipated demands need – by definition – to be forecasted. Forecasting is a huge discussion domain in itself. It is not the purpose of this text to have an extensive treatise. Yet some important guidelines can be formulated.

Forecasting is about gathering of data about future customers' behaviour. It is by definition a difficult challenge. Often customers do not themselves know what and when they will buy. How can we expect the sales organization to forecast their future purchases? The impossibility of forecasting may however be circumvented. Are forecasts really required at the level of finished products? At what level need forecasts to be available? This is one of the crucial questions that many production companies simply disregard.

An MTS company clearly needs forecasts at the level of finished products. Ultimately, products need to be produced to forecast, and therefore forecasts at that level need to be available.

ATO companies do however not need a single forecast at the level of finished products. Instead what is required is some visibility in terms of the product modules that will be required to support future demand. This information can often be made available starting from forecasts at product family level. The forecasted number of required modules can be derived by simple explosion using planning bills. Less commonly used modules may not necessarily be planned using the same approach but could be planned using a simple reorder policy.

MTO companies would – at first sight – not need any forecast. Yet these companies also need some visibility in order to make available the necessary productive capacity. Also, they may want to procure long lead time items to forecast. The same principle applies. Forecasting at product family level will surely do. Those forecasts can be exploded using planning bills to derive a forecasted number of required 'kits of parts'.

A second remark with respect to forecasting relates to forecasting accuracy. Planning need not be as accurate in the distant planning horizon as it should be in the nearby planning horizon. The demands on the forecasting accuracy are less stringent for the distant planning horizon. Where forecasts are typically required in weekly time buckets at the start of the planning horizon, monthly time buckets are often sufficient towards the end of the planning horizon.

Whatever the level of forecasting, both anticipated demands and firm demands need to be expressed in terms of the MPS items. As mentioned above, forecasts for product families are typically exploded using planning bills into forecast statements for the individual MPS items. But also firm demands need to be 'configured' expressed in terms of the MPS items. It then becomes possible to compare firm demands with forecasted demands and to develop a statement of net demand for each of the MPS items. The net demand is a statement of what management believes the future demand to be. It is often calculated using a selected forecast consumption rule. Forecast consumption will be discussed as part of the discussion on order promising.

4.3.5.6 Development of a tentative MPS

An MPS consists of a schedule of orders for each of the MPS items. Each schedule is typically developed in isolation of the other schedules, aiming at satisfying the corresponding demands, within the boundaries as expressed by the previous schedule and in accordance with any targets as set by the responsiveness policy.

Figure 4.26 illustrates an MPS at the start of week 1 and its development at the start of week 2. The provisioning time fence is 7 weeks out in the future and divides the horizon into the provisioning horizon and the tentative horizon. The provisioning horizon covers the cumulative lead time of the considered MPS item. As time progresses, the provisioning time fence is repositioned at the same distance in the future. The provisioning horizon moves along the time scale and progressively covers what previously has been part of the tentative horizon.

The tentative MPS is developed within the constrained availability of materials. Net demand has increased within the provisioning horizon. Yet, the corresponding MPS item orders are not increased automatically, since they are within the cumulative lead time of the MPS item. The net demand increase results in a change of the MPS order in week 8 (week 8 is just out the provisioning horizon). The change raises projected available inventory to 6 units at the end of the provisioning horizon, as may be aimed at by a predefined responsiveness policy. One time bucket is added to restore the original length of the planning horizon.

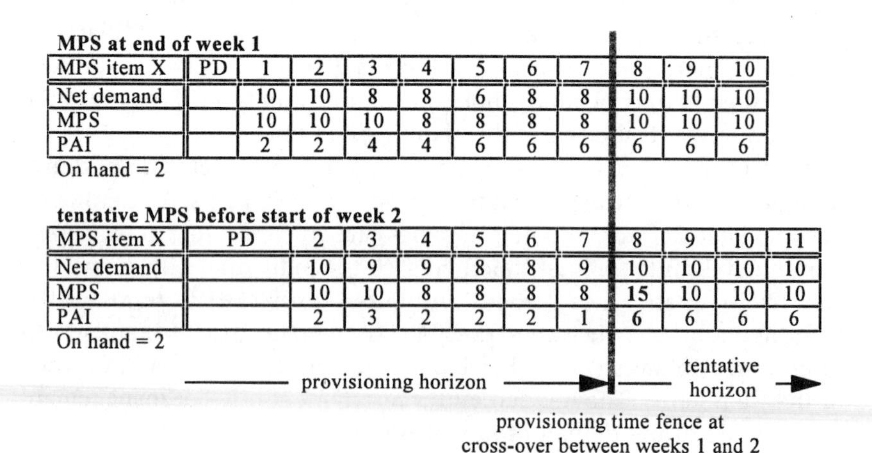

MPS at end of week 1

MPS item X	PD	1	2	3	4	5	6	7	8	9	10
Net demand		10	10	8	8	6	8	8	10	10	10
MPS		10	10	10	8	8	8	8	10	10	10
PAI		2	2	4	4	6	6	6	6	6	6

On hand = 2

tentative MPS before start of week 2

MPS item X	PD	2	3	4	5	6	7	8	9	10	11
Net demand		10	9	9	8	8	9	10	10	10	10
MPS		10	10	8	8	8	8	15	10	10	10
PAI		2	3	2	2	2	1	6	6	6	6

On hand = 2

Figure 4.26 Development of a tentative MPS.

Note that the tentative MPS is developed while taking account of materials considerations, but without taking account of capacity considerations. Tentative master production schedules are still to be evaluated in terms of their feasibility with respect to available productive capacity.

4.3.5.7 Rough-cut capacity planning

Objectives

Rough-cut capacity planning aims at evaluating a tentative MPS with respect to available productive capacity. The objective is to estimate the loads that would be placed by the tentative MPS on the company's resources, and to compare these loads with available productive capacity. Where the tentative MPS would appear to be overstated, management should modify the tentative MPS or take corrective actions to increase available capacity. Ideally, rough-cut capacity planning should ensure that MRP be driven by an MPS that is feasible to realize.

The role of rough-cut capacity planning is to be able to compare different tentative master production schedules, before running MRP. Any rough-cut capacity planning technique starts therefore from the tentative MPS, without considering the detailed output that would result if MRP were driven by the tentative MPS under consideration. Rough-cut capacity planning is a simple technique to avoid having to use MRP and the corresponding CRP technique (which will be discussed later) as 'what if?' simulators to test the feasibility of the tentative MPS.

Rough-cut capacity planning techniques are therefore really 'rough-cut' approaches. No account can be taken of the impact of work-in-process inventories, or of that of lot sizing, since the impact of these only becomes clear when finally running MRP. Rough-cut capacity planning techniques simply translate the tentative MPS to estimated loads. This translation is based upon standard loads attributed to each of the MPS items. The standard loads indicate how the production of one unit of each of the MPS items is expected to load the company's resources. The techniques assume that for the production of one unit of an MPS item, the MPS item itself is produced, as well as its constituent components and subassemblies.

Two rough-cut capacity planning techniques, the bill of labour and the resource profiles methods, will be discussed by means of a common example, which is illustrated in Figure 4.27.

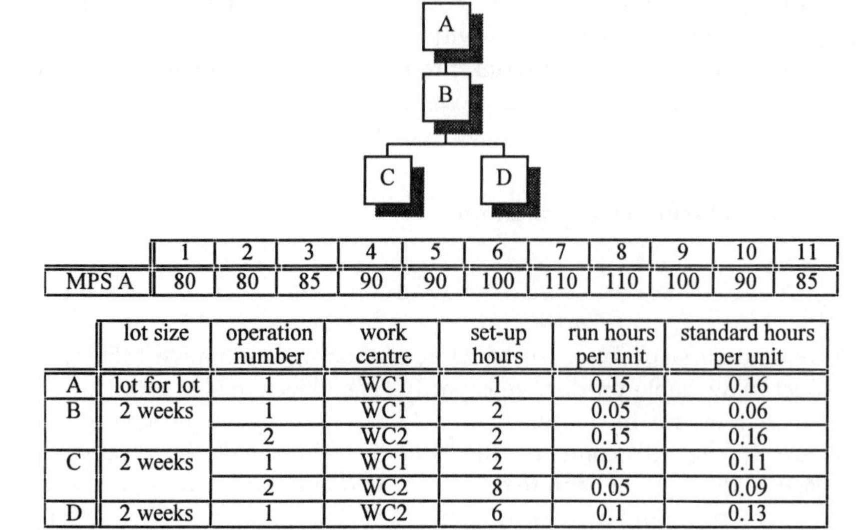

	1	2	3	4	5	6	7	8	9	10	11
MPS A	80	80	85	90	90	100	110	110	100	90	85

	lot size	operation number	work centre	set-up hours	run hours per unit	standard hours per unit
A	lot for lot	1	WC1	1	0.15	0.16
B	2 weeks	1	WC1	2	0.05	0.06
		2	WC2	2	0.15	0.16
C	2 weeks	1	WC1	2	0.1	0.11
		2	WC2	8	0.05	0.09
D	2 weeks	1	WC2	6	0.1	0.13

Figure 4.27 Bill of material, master production schedule and routeing for product A.

Figure 4.27 defines a (tentative) MPS for a product A, its bill of material as well as the corresponding routeing information. The timing convention as regards MPS quantities is that these should be made available at the start of the relevant time buckets. The routeing information lists the required operations for each conversion step in the bills of material, as well as the corresponding run time per unit, set-up time and standard time. Standard times will be the basis for deriving the standard loads for the MPS items. They are derived as the sum of the run time per unit and the average set-up time per unit. The average set-up time is derived from the lot size estimate. Lot sizing for product A is lot for lot. The average lot size is estimated as the average weekly demand, which is about 100 units. Lot sizing for items B, C and D is to provide for a 2-week period of supply. The average lot size is therefore estimated to be 200 units, the average demand across 2 weeks.

Bill of labour

The bill of labour approach is a first technique for rough-cut capacity planning. A bill of labour identifies the standard load of an MPS item on each of the (relevant) work centres. The standard load of an MPS item on a work centre is calculated as the sum of the standard loads on the

	load of MPS item A (hours per unit)	
WC1	0.16 + 0.06 + 0.11 =	0.33
WC2	0.16 + 0.09 + 0.13 =	0.38

Figure 4.28 Bill of labour for MPS item.

work centre of all the operations involved in the production of one unit of the MPS item, **including** its subassemblies and components. The calculation for MPS item A is illustrated in Figure 4.28.

The bills of labour are used to translate the master production schedules to load projections for each of the involved work centres. The translation is a simple matrix manipulation between the matrices of master production schedules and the bill of labour matrix (in case of multiple MPS items). The result for our example is represented in Figure 4.29. The capacity of work centres 1 and 2 is assumed to be 40 hours per week.

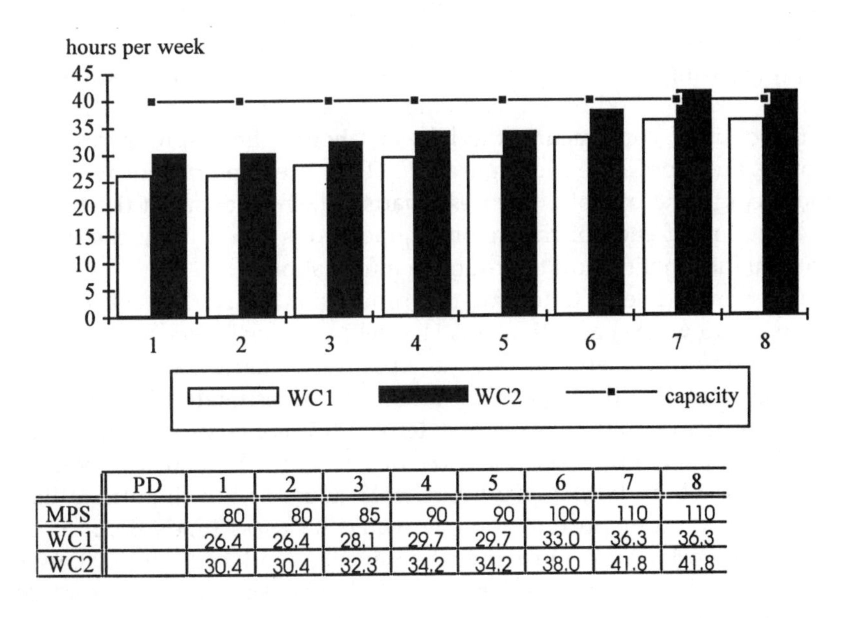

	PD	1	2	3	4	5	6	7	8
MPS		80	80	85	90	90	100	110	110
WC1		26.4	26.4	28.1	29.7	29.7	33.0	36.3	36.3
WC2		30.4	30.4	32.3	34.2	34.2	38.0	41.8	41.8

Figure 4.29 Capacity requirements according to RCCP using bills of labour.

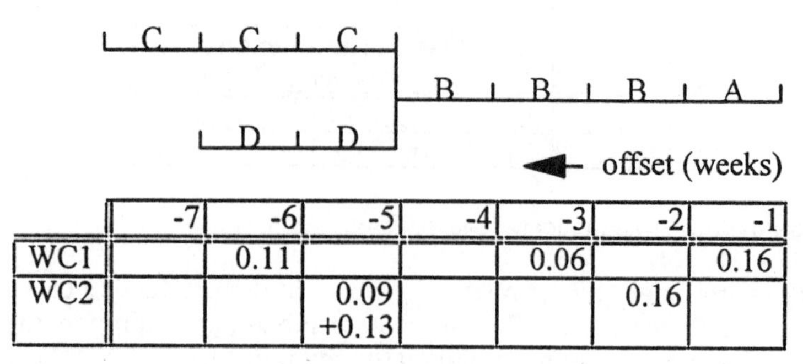

	-7	-6	-5	-4	-3	-2	-1
WC1		0.11			0.06		0.16
WC2			0.09 +0.13			0.16	

Figure 4.30 Operation process chart and resource profile for product A.

Resource profiles

The bills of labour approach does not take into account the fact that operations and the corresponding loads are offset from each other. The approach finds slight overloads for work centre 2 in weeks 7 and 8, in accordance with the higher MPS quantities in those weeks, although the execution of the operations is offset from the timing of the MPS quantities as a result of operation lead times. Rough-cut capacity planning using resource profiles tries to overcome the shortcomings of the previous method by actually time-phasing the information that was at the basis of constructing the bills of labour. The resulting resource profiles are represented in Figure 4.30. They are constructed by taking into account the operation process charts, which reconstruct the timing of execution of each of the involved operations based upon the bill of material information and the routeing information.

The planned lead times for the production of D (2 weeks), C (3 weeks), B (3 weeks) and finally A (1 week) were estimated based upon the routeing information as well as upon estimated queuing times for work centres 1 and 2. Queuing times can be estimated based upon (again) routeing information, or be derived from observed queue length statistics. In our example, a lot size of 100 units A on work centre 1 would take about 0.15 × 100 hours = 15 hours of run time (note that other typical orders in our example would take less time). So, if a job arrived at work centre 1 just after the start of a work order for 100 units A, it would have to queue for about 15 hours. The queue time of 20 hours is therefore a very comfortable estimate. Likewise, the maximum run time for a typical work order on work centre B can be estimated to be 30 hours (a lot size of 200 units B at a run time of 0.15 hours per unit). The queuing time has correspondingly been set at 30 hours. Planned lead time estimates derive from the actual routeing information, while taking account of estimated queue lengths. For instance, a work order of 200 units B would take 20 hours (queue before work centre 1) + 2 hours (set-up on work centre 1) + 0.05 × 200 (run time on work centre 1) + 30 hours (queue before work centre 2) + 2 hours (set-up of work centre 2) + 0.15 × 200 (run time on work centre 2) which is about 94 hours. The overall lead time for MRP was therefore set at 3 weeks. This means that the planned order release of a work order for B will be offset from the planned order receipt by a lead time of 3 weeks. These 3 weeks should be sufficient to cover queuing times before work stations, as well as actual set-up times and run times. Note that actual MRP systems take account of other lead times as well (such as wait and move lead times), but these would not really impact on the reasoning within the example.

This example assumes that MPS quantities should be made available at the start of each time bucket. The production of item A therefore loads work centre 1 in the preceding time bucket (an offset of −1 period with respect to the planned MPS quantity). The production of item B takes three periods, but loads would be backwards scheduled. When taking account of run, set-up and queue times, work centre 2 would be loaded in the last available time bucket (offset −2 periods with respect to the planned MPS quantity), and work centre 1 in the preceding one (offset −3 periods). The first of the three available time buckets to produce item B covers any surplus flexibility in lead time and of course the queuing time in front of work centre 1. The same logic applies to the production of items C and D.

The result of the resource profile approach is represented in Figure 4.31. The past due loads should under normal conditions already have been completed (therefore they are not part of the graphical representation). Note that the peak load for work centre 2 has moved forward to weeks 2–5.

Capacity requirements versus capacity available

Rough-cut capacity planning provides a projection of load for the tentative MPS under consideration. This projection of load has to be compared with available productive capacity. Under the assumption of a more or less reliable rough-cut load projection, traditional thinking in the domain of rough-cut capacity planning assumes that a work centre is not overloaded when the capacity requirements are below or equal to capacity available. Capacity available is traditionally defined as the maximum output under normal working conditions.

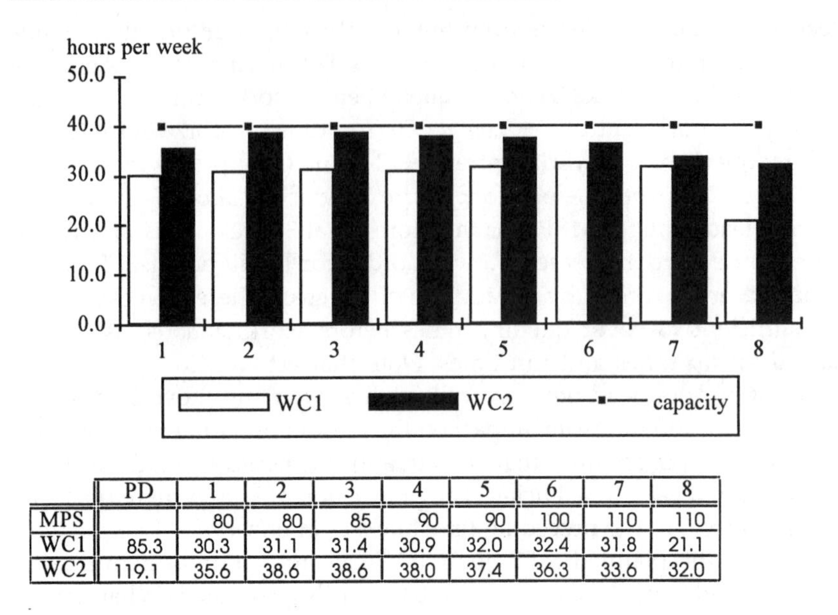

	PD	1	2	3	4	5	6	7	8
MPS		80	80	85	90	90	100	110	110
WC1	85.3	30.3	31.1	31.4	30.9	32.0	32.4	31.8	21.1
WC2	119.1	35.6	38.6	38.6	38.0	37.4	36.3	33.6	32.0

Figure 4.31 Capacity requirements according to RCCP using resource profile.

Recent books and theories have provided more insight in the theory of capacity management and have demonstrated that the above comparison of capacity requirements versus capacity available to assess the feasibility of a tentative MPS is oversimplified.

Goldratt was the first to point out not only the importance of the identification of bottlenecks, but also the importance of how these are scheduled (see in particular Goldratt and Cox 1986, Goldratt 1990, Umble and Srikanth 1990). A production programme resulting in bottlenecks with capacity requirements equal to capacity availability is only feasible if the bottlenecks can be guaranteed sufficient input so that they never fall short of work. In his words, 'any hour lost on a bottleneck resource is an hour lost for the entire production system' and makes the original production programme unfeasible. The contribution was important since it made explicit the fact that work centres in a production environment are dependent upon each other. Lack of synchronization between bottleneck resources and their feeding resources leads to loss of productive capacity and invalidates the production programme.

Goldratt's ideas were further developed in what came to be called the theory of synchronous manufacturing. This theory explains that even non-bottleneck resources may endanger the realization of an MPS, and starts from two basic phenomena to explain the complexities of the production environment. The first phenomenon is the actual existence of dependencies. The executions of planned operations are determined by the completion of preceding operations at other work centres and priority setting of different planned operations queuing at a work centre affects the synchronization with downstream operations. A second phenomenon is the existence of random events and statistical fluctuations. Random events are any kind of events that cannot be anticipated but have a disruptive effect on the flow of materials. Statistical fluctuations are those fluctuations caused by the inherent variability of any process. It is the existence of both phenomena that makes that certain resources, although not bottleneck resources and hence with capacity requirements below capacity availability, temporarily constrain production output and potentially invalidate the realization of the MPS. Such resources are called capacity constraint resources. They have to be scheduled carefully if production output is to be guaranteed.

Blackstone (1989) looks at the same problem from another point of view. Blackstone looks at work centres in a production system as queuing systems. Queuing systems are characterized by queues in front of them, which shrink and grow as a result of varying input and output. Input rates are determined by completion times at upstream work centres. Output rates are determined by work centre utilization and efficiency. Even with average input below or equal to average output, i.e. with capacity requirements below or equal to capacity availability, queues may grow as a result of input bursts that temporarily exceed the processing capability of the queuing system. These queues are reduced in size during time periods when output exceeds input. The queue may even completely dry up, as a result of continuous processing with a temporary shortfall of input. If capacity requirements are equal or nearly equal to available productive capacity, any temporary shortfall of input implies lost productive capacity. Loss of productive capacity endangers the realization of the MPS. Also queues grow, which further aggravates the problem of synchronization among different work centres. If MRP is to work reliably, the length of queues should be within the planned lead times as defined for the different operations.

Systems behave like queuing systems as soon as input and output rates behave like independent variables. As soon as the input is controlled such that the system never falls short of work, loss of productive capacity can be avoided and queues can be controlled. This is precisely the objective of the synchronous manufacturing theory, i.e. to schedule the capacity constraint resources and gear the feeding work centres with these schedules so that production flow can be synchronized, loss of productive capacity can be avoided and the proposed production programme can be realized. The concepts of the theory of synchronous manufacturing are clear. It is a radical new way of thinking, which may result in a new approach for production planning that deviates from the MRP II-based production planning approach. Its development looks promising but implementation appears to be difficult. Meanwhile, many companies want to capitalize on their major investments in their MRP II-based planning architecture. They want to fine tune current MRP II-based thinking by taking into account new emerging ideas.

The MRP II-based planning approach starts by developing the MPS, before using the MPS as driver of MRP. Material requirements planning is a straightforward calculation logic that plans procurement and production in support of the MPS using a combination of planned lead times, lot sizing policies, safety stock parameters, etc. The focus of MRP is on planning materials, not on developing schedules for resources that temporarily could constrain production output. Against this background and with the existence of dependencies, random events and statistical fluctuations, the major conclusion must be that capacity requirements below but nearly equal to capacity available are not sufficient to guarantee feasibility of the MPS. Or, as Blackstone puts it: 'as a general rule, capacity available must be greater than capacity required in order to avoid the problems of queues that grow indefinitely'. No general guidelines can however be given as regards the required surplus capacity. The required 'underloading' of a work centre to keep queues within predetermined lead times is dependent upon such factors as the size of time buckets within master production scheduling, the degree of lot sizing within MRP, process reliability, the way lead times were determined, the actual mix within the MPS and the quality of the rough-cut capacity plan.

Load, available capacity, unit of measure

Rough-cut capacity planning can be made simple in practice. More 'natural' units of measure can be used to express available productive capacity and corresponding loads. Examples are number of cars, tonnes of finished product, number of batches, even pairs of shoes. These measures are easy to understand and interpret, and simple to calculate. Moreover, they often have a meaning from a sales point of view, which provides a better basis for interpretation.

4.3.5.8 Capacity requirements planning

Objectives

CRP is a specific technique that aims at a deterministic simulation of capacity requirements using the output of the MRP explosion. CRP evaluates all open orders and orders planned by MRP and therefore takes into account work in progress, lot sizing policies and time phasing. CRP resolves the shortcomings of rough-cut capacity planning, but its information can only be made available after laborious calculations based upon the MRP output. MRP itself is capacity insensitive. The role of CRP is to act as a final check from a capacity point of view.

Development of a capacity requirements plan

The development of the capacity requirements plan is illustrated by means of the same example that has been used for purposes of discussion of rough-cut capacity planning (Figure 4.27). Figure 4.32 represents the MPS for product A and the material requirements plans for items B, C and D. Open orders are indicated as scheduled receipts. The timing conventions are that scheduled receipts are scheduled to be received at the start of the relevant time buckets. Operations should be or have been executed in the preceding time buckets. Planned order releases are released at the start of the action time bucket, e.g. the first time bucket of the planning horizon. Capacity requirements planning calculates capacity requirements corresponding with both planned and open orders. They are typically calculated backwards with reference to the scheduled or planned receipt dates, respectively, for open and planned orders.

The lead times within the CRP run are not estimated as in rough-cut capacity planning or in MRP, but calculated. The lead time for a certain operation is derived as the run time per unit times the actual lot size, incremented with set-up and queue time. Be aware that the queue time parameters are still guess parameters within the lead time calculation. The queue times represent the average time jobs have to wait before being processed on the corresponding work centre. These may actually heavily depend on the degree of utilization of the work centres. As in the RCCP calculations, queue times are estimated to be 20 hours for work centre 1 and 30 hours for work centre 2.

The calculated lead times serve to offset the loads from the scheduled or planned receipt dates. Loads are typically spread across the complete run time, rather than positioned at, for instance, the start or end of the run operation.

		1	2	3	4	5	6	7	8	9
A	MPS	80	80	85	90	90	100	110	110	100
	Scheduled receipts	75								
	PAI	5	20							
	Planned order receipts		95	85	90	90	100	110	110	100
	Planned order releases	95	85	90	90	100	110	110	100	90

On hand = 10; lot for lot; safety stock = 20

		1	2	3	4	5	6	7	8	9
B	Gross requirements	95	85	90	90	100	110	110	100	90
	Scheduled receipts		200							
	PAI	5	120	30	100	0	110	0	90	0
	Planned order receipts				160		220		190	
	Planned order releases	160		220		190		165		

On hand = 100; 2 weeks supply lot sizing

		1	2	3	4	5	6	7	8
C	Gross requirements	160		220		190		165	
	Scheduled receipts	180		210					
	PAI	20	20	10	10				
	Planned order receipts					180		165	
	Planned order releases		180		165				

On hand = 0; 2 weeks supply lot sizing

		1	2	3	4	5	6	7	8
D	Gross requirements	160		220		190		165	
	Scheduled receipts	180							
	PAI	30	30						
	Planned order receipts			190		190		165	
	Planned order releases	190		190		165			

On hand = 10; 2 weeks supply lot sizing

Figure 4.32 Material requirements plan.

The capacity requirements for the planned orders are represented in Figure 4.33. Figures in italic represent scheduled receipt quantities and their corresponding loads. Part of the required operations may already have been completed. MRP II is typically informed about these completed operations by a shop floor control and feedback module.

Lead times and loads are calculated for each individual open or planned work order. The scheduled receipt of 200 units B at the start of week 2 would load work centre 2 for 30 hours (0.15 × 200) run time and 2 hours set-up time. The example assumes that operation 1 has been completed, so work centre 1 is not loaded. The planned receipt of 160 units B at the start of week 4 would load work centre 2 for 24 hours (0.15 × 160) run time and 2 hours set-up time. The expected completion date for operation 1 on work centre 1 is offset from the expected start date for operation 2 by the queuing time for work centre 2, which is 30 hours. This implies that operation 1 would load work centre 1 during week 2

	PD	1	2	3	4	5	6	7	
A: order receipts		***75***	**95**	**85**	**90**	**90**	**100**	**110**	
set-up hours WC1		1	1	1	1	1	1		
run hours WC1		14.25	12.75	13.5	13.5	15	16.5		
B: order receipts			***200***		**160**		**220**		**190**
set-up hours WC1			2		2		2		
run hours WC1			8		11		9.5		
set-up hours WC2		*2*		2		2		2	
run hours WC2		*30*		24		33		28.5	
C: order receipts		***180***		***210***		**180**		**165**	
set-up hours WC1		2		2		2			
run hours WC1		21		18		16.5			
set-up hours WC2	*8*		8		8		8		
run hours WC2	*9*		10.5		9		8.25		
D: order receipts		***180***		**190**		**190**		**165**	
set-up hours WC1			6		6		6		
run hours WC1			19		19		16.5		
Total load WC1		38.25	23.75	34.5	27.5	34.5	29		
Total load WC2	17	32	43.5	26	42	35	38.75		
Cum. total load WC1		38.25	62	96.5	124	158.5	187.5		
Cum. total load WC2		49	92.5	118.5	160.5	195.5	234.3		

Figure 4.33 Work centre loads of scheduled and planned order receipts.

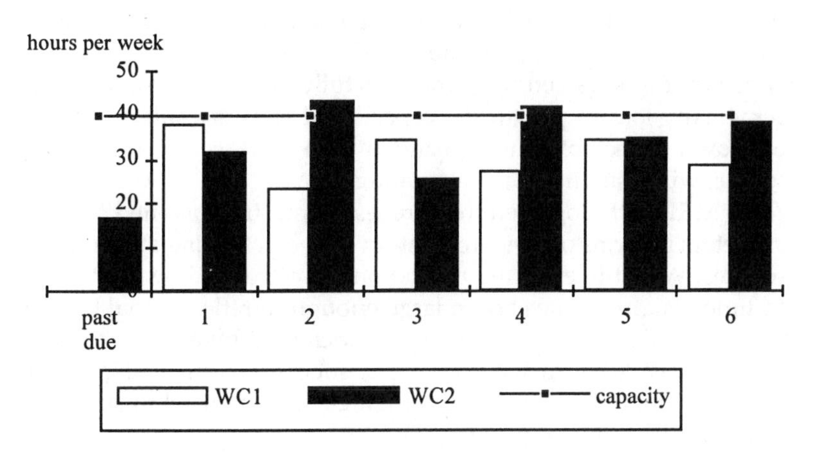

Figure 4.34 Capacity requirements according to CRP.

for 8 hours (0.05 × 160) run time and 2 hours set-up time. The same logic applies to the rest of the scheduled and planned receipts. Note however that operation 2 of the work order of 180 units C scheduled to be received at the start of week 1 has not been confirmed yet, resulting in a past due load on work centre 2.

Total capacity requirements are illustrated graphically in Figure 4.34, and cumulatively by Figure 4.35.

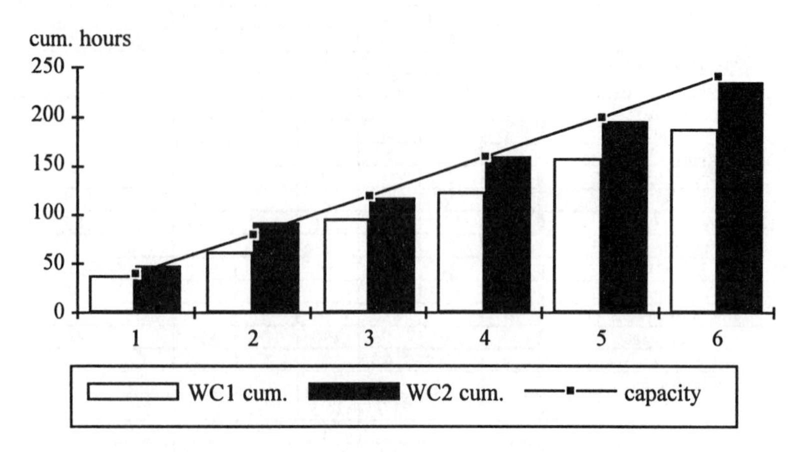

Figure 4.35 Cumulative capacity requirements according to CRP.

The cumulative view emphasizes the critical balance between available capacity and capacity requirements for work centre 2. Results should be preferably be looked at in a cumulative way, since this allows the room for manoeuvre to been seen more easily. There is nearly no room for manoeuvre for work centre 2, since it is fully loaded up to and including week 4. Any loss of productive capacity, be it as a result of temporary breakdown or lack of synchronization with work centre 1 as a feeding resource, will result in growing queuing times.

The MRP/CRP logic entirely relies upon the planned lead time parameters in terms of synchronization between feeding and consuming resources. With the expected utilizations as calculated by CRP, planned lead times might not be chosen large enough to reflect actual lead times on the shop floor. Problems may be anticipated by a series of possible measures (overtime on work centre 2, subcontracting, order splitting). Otherwise, finite capacity scheduling logic may be used to simulate the proposed MRP scenario (Chapter 5).

4.3.5.9 Evaluation

The current approach for development of an MPS is very top down in nature. A tentative MPS is developed first before running rough-cut capacity planning. Only then are MRP and CRP run to plan all other items and to perform a final deterministic capacity check.

Two RCCP techniques were discussed. The bills of labour approach is a simple technique, which however should be used within the limits of its applicability. Time phasing is not taken into account, and results should therefore best be aggregated in time buckets that are quite large with respect to total cumulative lead time. The presentation of data using small time buckets creates a wrong impression of accuracy.

The resource profiles approach adds time phasing to the picture, with the purpose of generating a more accurate load picture. The approach, however, remains limited in its endeavour to reach accuracy. It is still rough cut in nature and cannot take into account work-in-progress inventories or lot sizing policies at lower levels of the bills of material. Lot sizing is actually quite a common approach. The theory of synchronized manufacturing advocates the use of lot sizing at overloaded work centres to save set-up time and increase throughput. The resource profiles approach requires a lot more computational effort, and will generate extra useful information only in the case of long production lead times and when the impact of lot sizing is limited.

The load projection as generated by CRP clearly deviates from the ones generated by the rough-cut capacity planning approaches. Capacity requirements planning is based upon the output of MRP. It is therefore sensitive to lot sizing, work-in-progress inventories and arrears, and is a superior capacity planning technique to the rough-cut capacity planning approaches described before. The MRP and CRP calculations are, however, so laborious that most companies do not use CRP as a simulation tool during development of the MPS. Anyhow, even if MRP and CRP could be run fast enough, their applicability as 'what if?' simulators during development of the MPS would be limited. MRP is a simple calculation logic that does not take into account finite capacity of production resources. CRP detects overloads but does not help to try to resolve those. CRP therefore is used as a mechanism to anticipate future resource requirements and fine tune capacity availability to meet temporary overloads.

In summary, rough-cut capacity planning is sufficiently good in evaluating the long-term feasibility of the plan. CRP should preferably be used to fine-tune capacity available to capacity requirements over the short horizon based upon detailed MRP output.

During the discussion of rough-cut capacity planning, it was made clear that some surplus capacity must be available to account for a potential lack of synchronization between different work centres. Work centres may lose productive capacity as a result of a temporary shortfall of input. This problem is especially relevant for rough-cut capacity planning, since the time buckets used for visualization of the load projections are usually large with respect to production lead times. The rough-cut load projections are only estimates of future load, expressed by month or quarter. The estimates may turn out to be load averages, which exhibit important fluctuations when looked at on a weekly basis. It should be clear that the same holds true for the interpretation of CRP results. CRP is not a finite capacity scheduler and just relies upon calculated lead times to generate a load projection for each of the work centres based upon the detailed MRP output. The MRP planned orders have not been sequenced yet, and a lack of synchronization between successive work centres may need to be compensated by some surplus capacity. Available capacity equal to capacity requirements is only acceptable to the extent that the bottleneck resource can be guaranteed sufficient input so that it never falls short of work. The MRP functionality has difficulty in guaranteeing this requirement. The only approach would be to make planned lead times sufficiently long so that enough work is released in advance to create a sufficiently large buffer in front of the bottleneck resource.

It must be stressed that companies successfully working with the MRP-based planning architecture are **primarily** faced with the management of materials. Their prime capacity constraints are often within the realization or final assembly stage of production. The investment stage of production often has some excess capacity to be able to cope with changing loads corresponding with changing mixes of demand (typical for capital goods and durables industry sectors). A capacity check for these companies is often easy to perform, for instance by checking the MPS against the throughput capacity within the realization stage. This capacity can be expressed in number of products per week, or any other measure.

Process industries, on the other hand, are primarily faced with capacity constraints throughout the investment and realization stages of production. These industries are less faced with the complexities of raw materials supply. The number of raw materials is often limited and bulk supply is common. Process industries cannot find in MRP II an ideal planning framework. Rather than to develop material plans that are checked against capacity, their objective is more to develop capacity plans that are checked against material availability. This topic will be further developed in section 4.4.

4.3.6 RESPONDING TO DEMANDS THROUGH MASTER PLANNING

4.3.6.1 Introduction

The primary role of master planning is quite naturally to drive the production system in response to firm and anticipated demands.

Equally important, however, is the ability to determine delivery dates for new or changed demands. This constitutes the second role of master planning, i.e. order promising. In this section, order promising within the MRP II-based planning and control architecture is dealt with.

Order promising should quite obviously be based upon the planning constraints that the company is faced with. The constraints determine the short- and long-term production capabilities, and as a result the boundaries for order acceptance. Planning constraints may either be material or capacity constraints. Material constraints are those related with the limited availability of material in the supply chain. Material constraints are typically short to medium term in nature. They are limited to the provisioning horizons, which are determined by the cumulative lead times for each of the MPS items. Capacity constraints are a result of limited productive capacity, both short term and long term.

The MPS drives the investment operations and represents the response of the production system to both firm and anticipated demands. It can and should therefore be seen as a good representation of the constraints facing the company. On the one hand, it reflects the amounts of material in the supply chain which have been purchased and/or produced to accommodate a certain demand picture (the schedule for each MPS item corresponds with investments in the form of released purchase and production orders). On the other, the schedule is a feasible schedule with respect to available productive capacity. It is therefore quite obvious that the MPS should be used as the basis for order promising.

Order promising is particularly relevant in ATO environments, in which the final assembly of a finished product is dependent upon availability of its constituent modules, made available by the MPS.

Order promising is equally relevant in MTS environments, when considering not the end customer but the distribution organization as the true customer of the production system. Replenishment orders are the real demands for the production system. A formal order promising system between production and distribution may help clarify responsibilities at both sides. Production may – despite the MTS environment – even try to improve its service level towards distribution by modularizing products and adopting an ATO planning approach. Product family forecasts are then sufficient to drive production beyond the final assembly horizon, whereas firm commitments as regards the mix of products are required at the start of final assembly.

In an MTO company, there is no real MPS, only a realization schedule, which reflects the promised customer orders. Some master schedules may be developed however for kits of parts, in case customer lead times are compressed by purchasing part of the materials to forecast. These master schedules are similar in nature to the master production schedules in an ATO or MTS environment. They drive the investment operations within the production system, and determine the amounts of material in the supply chain. In case such master schedules exist, they should be used as the basis for order promising, together with supplementary information on available productive capacity. Order promising in an MTO environment is to a large extent about allocating production capacity to incoming demands.

4.3.6.2 The available-to-promise logic

The available-to-promise (ATP) logic is used to support the order promising process. ATP quantities are quantities of MPS items that are scheduled as part of the MPS, but have not yet been allocated to specific demands and are still available to satisfy new demands or demand configuration changes.

The ATP quantities are calculated from the information in the MPS planning tables. Figure 4.36 illustrates the concept of ATP.

MPS item X	1	2	3	4	5
Forecast	10	10	10	10	10
Customer orders	8	8	4	4	
Net demand	10	10	10	10	10
MPS	25		25		10
PAI	20	10	25	15	15
Available-to-promise	14		17		10

On hand = 5

Figure 4.36 Available to promise.

The example shows the MPS planning table for a finished product X, which may be promised individually without taking into account the availability of other MPS items. Total demand is forecasted at 10 units per week. Firm orders have been received totalling 8 units in weeks 1 and 2 and 4 units in weeks 3 and 4. The net demand expresses the demand imposed on the production system, taking into account forecast and firm orders. Its relevance will become clear when discussing the concepts of forecast consumption. At this stage, net demand equates total expected demand or forecast. The MPS is developed as the response of the production system to the imposed demands. The example shows three firm planned orders, resulting in the indicated projected available inventory profile. The **available to promise** entry indicates what is currently projected to be available for allocation. ATP quantities are calculated by comparing the supply (available on hand and the MPS) against that which has already been promised. The timing convention for the ATP quantities corresponds with the one for the MPS quantities. If MPS quantities are planned to be available at the start of the corresponding time buckets, then the ATP quantities are also available at the start of the corresponding time buckets. The ATP quantity for week 1 is for instance calculated as (5 units on hand) + (25 units supply) – (8 units allocated in week 1) – (8 units allocated in week 2) = (14 units available as of start of week 1). The same calculation logic applies for the other ATP quantities. The ATP entry is the basis for order promising. Consider that an order of 20 units arrives to be delivered in week 5. These 20 units can be promised since the cumulative availability for allocation in week 5 is determined as (14 units available as of start of week 1) + (17 units available as of start of week 3) + (10 units available as of start of week 5) = (41 units available for allocation within week 5). Since the purpose should be to keep as much flexibility and produce as much 'just-in-time' as possible, 10 units are promised out of the MPS quantity in week 5, and 10 extra units out of the MPS quantity in week 3. The updated planning table is represented by Figure 4.37. Forecasts have not yet been changed to reflect the incoming order.

MPS item X	1	2	3	4	5
Forecast	10	10	10	10	10
Customer orders	8	8	4	4	20
Net demand	10	10	10	10	20
MPS	25		25		10
PAI	20	10	25	15	5
Available-to-promise	14		7		

On hand = 5

Figure 4.37 Figure 4.36 after promising of extra demand.

The ATP logic is a straightforward mechanism to keep track of unallocated MPS quantities. Order promising should entirely be based upon this information. Situations may however be more complex than the one illustrated above. Order promising may involve promising of different MPS items simultaneously to satisfy a certain demand, particularly in ATO environments, where different modules are required to assemble the desired product or set of products. Order promising then involves two basic steps, which are the configuration stage and the allocation stage. The configuration stage involves configuring the incoming demand in terms of the corresponding MPS item requirements. The allocation stage involves the determination of a delivery date, while taking account of the requested delivery date and the required ATP quantities. Upon acceptance of the proposed delivery date by the customer, the required MPS quantities are allocated to the concerned demand, and ATP is correspondingly updated.

Consider the PC example of Figure 4.15 with planning tables for two of the MPS items, as in Figure 4.38. During the configuration stage, any incoming demand for a tower frame PC would call for one MPS item 'common items', as well as one MPS item 'tower frame'. These MPS requirements need to be offset from requested delivery date to account for the final assembly lead time. The second step involves the evaluation whether all MPS requirements can be satisfied. With the status of Figure 4.38, a maximum of 27 PCs of which 13 tower frame PCs could be promised for delivery in week 5 (MPS requirements in week 3 with final assembly lead time of 2 weeks).

Common items	1	2	3	4	5
Forecast	10	10	10	10	10
Customer orders	8	6	4		
Net demand	10	10	10	10	20
MPS	20		20		20
PAI	15	5	15	5	15
Available-to-promise	11		16		20
Cumulative ATP	11	11	27	27	47

On hand = 5

Tower frame	1	2	3	4	5
Forecast	5	5	5	5	5
Customer orders	3	1			
Net demand	5	5	5	5	5
MPS		10		10	
PAI	2	7	2	7	2
Available-to-promise	4	9		10	
Cumulative ATP	4	13	13	23	23

On hand = 7

Figure 4.38 Planning tables for two PC modules.

Order promising is part of demand management, which should aim at a real-time interface with the customer. This clearly puts in perspective the need for real-time maintenance of orders, as well as of ATP information.

4.3.6.3 Forecast consumption

The forecast expresses the total of firm and anticipated demands. As firm orders come in, they actually consume the forecast. Orders may consume forecast in many different ways, depending upon management's view on how to compare incoming orders with forecast. Some of the possible forecast consumption approaches can be formalized through forecast consumption rules. The forecast consumption rules determine how the net demand is arrived at, based upon the forecast and order information. Net demand expresses what management believes to be the future real demand on the production system.

The following discussion presents some of the possible forecast consumption rules. It must be anticipated however that this discussion is not exhaustive and the specific forecast consumption approach very much depends upon the particular relationship between the production system and its market.

Figures 4.36 and 4.37 have illustrated the principles of the simplest forecast consumption rule there is. Orders consume the forecasts, time bucket by time bucket. If orders exceed forecasts (such as for time bucket 5 of Figure 4.37), net demand is taken as the orders information. This results in a cumulative net increase of original net demand.

Other forecast consumption rules aim at keeping net demand stable, even when orders come in differently from originally expected. 'Overconsumption' in some time buckets is compensated by 'underconsumption' in other ones. These forecast consumption rules translate management's confidence in the total forecast quantities. A first approach is the backward forecast consumption. Backward forecast consumption is the process whereby if orders in one bucket exceed the forecast, then the excess is carried back and consumes unconsumed forecast in previous buckets. If there is insufficient unconsumed forecast, the net result is the increase of the net demand by the shortfall. An illustration is given in Figure 4.39. The overconsumption of 10 units in time bucket 5 is compensated by borrowing 6 units (the unconsumed forecast) from bucket 4 and a remainder of 4 units from bucket 3. Net demands are calculated accordingly. The net demand for buckets 3 and 4 equals the original forecast minus the borrowed quantities. The result is that total net demand equals total forecast. The overconsumption in time bucket 5 is attributed to anticipated demands that were realized later than anticipated.

	1	2	3	4	5
Forecast	10	10	10	10	10
Customer orders	8	8	4	4	20
Net demand	10	10	6	4	20

Figure 4.39 Backward forecast consumption.

Forward forecast consumption is similar to backward forecast consumption but overconsumption is borrowed from future time buckets. An illustration is given in Figure 4.40. The overconsumption in time bucket 2 is attributed to anticipated demands that were realized earlier than anticipated.

	1	2	3	4	5
Forecast	10	10	10	10	10
Customer orders	8	20	4	4	
Net demand	10	20	4	6	10

Figure 4.40 Forward forecast consumption.

The combined backward/forward forecast consumption overcomes the problems that may arise in the case of backward forecast consumption where not all of the excess can be borrowed from previous time buckets. The excess is first carried backward up to the limit of the available unconsumed forecast and the balance carried forward. An illustration is given in Figure 4.41.

	1	2	3	4	5
Forecast	10	10	10	10	10
Customer orders	8	20	4	4	
Net demand	8	20	4	8	10

Figure 4.41 Backward/foreward forecast consumption.

The development of net demand is important since it expresses what management believes to be the future demand. The MPS is developed in response to the net demand, and as such determines projected available inventory and available to promise. The updated available to promise projection then becomes the basis for promising further demands. Net demand, if wrongly assessed, may result in overstated master production schedules or master production schedules being unable to respond satisfactorily to incoming demands.

It is worthwhile at this stage commenting on two exception statuses of the MPS. A first exception status, 'overcommitment', occurs if the MPS cannot satisfy net demand, i.e. the total of firm and anticipated demands. This exception status is characterized by negative PAI figures. Appropriate order promising should ensure that orders are promised not as anticipated but within the boundaries of the available-to-promise information. The second exception status, 'overselling', occurs when more demands have been promised than the MPS can satisfy. This exception status is characterized by negative available-to-promise figures. This situation is alarming to the extent that some demands cannot be satisfied on time (if production cannot respond appropriately) and must be delayed to later time buckets when sufficient supply exists.

4.3.7 REQUIREMENTS OF A MASTER PLANNING SYSTEM

The discussion so far has highlighted two major functions within any master planning system: master production scheduling and demand management. These functions correspond to two major tasks that the master planning system has to perform as an interface between the production system and the market. On the one hand, an MPS should be developed to drive the production system. On the other, the MPS should be used as the basis for order promising. These two tasks have different characteristics.

An MPS should ideally remain valid over a certain period of time in order to enforce the desired degree of stability within the production system. It may be developed with a certain frequency (depending upon the bucket size) or its development may be triggered by certain disruptive events within the production system overhauling the existing MPS, or events in the marketplace requiring a thorough revision of forecasts.

In contrast, the company should respond to demands continuously and immediately. Order promising should be possible in real time by tracking promised demands and maintaining corresponding MPS item requirements and ATP information.

The different characteristics of the two primary tasks of master planning should be reflected by the master planning systems. The response of the master planning system towards the production system, characterized by a certain degree of inertia, should be triggered by progression of time or events. Its response towards the market, characterized by variability, should be real time. These characteristics should be satisfied whatever the type of production environment. The outline overviews of master planning systems for some possible types of production environment will now be discussed.

4.3.7.1 Master planning for MTS environments

MTS environments may be considered the less complex environments from a master planning point of view. The major building blocks of a master planning system for an MTS environment are represented in Figure 4.42.

Demand management consists of two primary functions, which are order promising and forecasting. The forecasting function identifies all actual and likely demands and compares actual demands with total expected demands to derive net demand. Net demand expresses what management believes to be the future demand imposed on the production system. Net demand is expressed as a schedule of quantities, by product. Net demand as well as the on-hand availability of finished products serves as the basis to develop the MPS. Forecasting (including the calculation of net demand) and the development of the MPS should be done with the same frequency as that of running MRP. MRP is usually run with a predetermined cycle time, e.g. weekly, or may exceptionally be run as a result of events that require the MPS to be reassessed. Forecasting and MPS development are therefore functions that are triggered either by progression of time or by certain events.

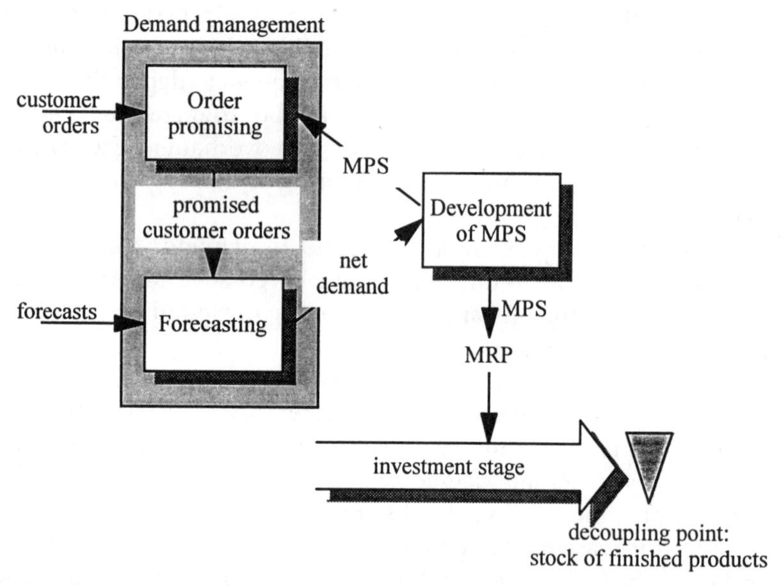

Figure 4.42 Master planning for an MTS environment.

Order promising should, in contrast, be done in real time. Order promising involves the determination of delivery dates for incoming demands. The demands are those placed upon the production system, and may be replenishment orders for the warehouses managed by the distribution organization. The required input for order promising is the MPS, as well as the on-hand availability of finished products. The order promising function should continuously compare on-hand availability and MPS with promised orders in order to have up-to-date ATP information.

4.3.7.2 Master planning for ATO environments

Figure 4.43 represents the architecture of a master planning system for the ATO environment. Two schedules are involved. The final assembly schedule responds to the customer orders promised over the final assembly horizon and drives the realization stage. The investment stage is driven by the MPS, which responds to the anticipated demand beyond the final assembly horizon (as expressed by the net demand), as well as to the MPS item requirements derived from the final assembly schedule. Note that MRP as a process is mentioned twice, firstly to derive MRP item requirements from the FAS schedule and secondly to explode the MPS to drive the investment stage of production. It should be clear that the 'size' of the first MRP explosion process cannot be compared with that of the second one. The bills of material defining end items starting from the MPS items are often quite flat. Lot sizing in the realization stage is also often limited. The translation of the FAS into MPS item requirements should actually be done in real time, so that the impact of any change of the FAS is equally known in terms of MPS item requirements. As is explained a bit further on, this is a necessary requirement to keep up-to-date ATP information that is maintained at the MPS item level.

The net demand is expressed in terms of the MPS items. It can be calculated in different ways. A 'logical' approach would be to calculate the net demand for each of the separate MPS items by comparing the MPS item forecast against the MPS item requirements as called for by the incoming customer orders. This approach is similar to the one for the MTS environment. Each MPS item is taken separately and a net demand statement is developed by comparing forecast requirements against customer order requirements. Forecasts for the individual MPS items can be made available by exploding product family forecasts through planning bills.

The alternative approach differs from the one described above in the way that forecast requirements are only compared with incoming customer orders at the level of the product family, and not for each of the individual MPS items. The net demand calculation at the level of the product family serves to derive 'unconsumed forecast', which is then exploded using planning bills to production forecasts (in other words 'dependent forecasts') for each of the individual MPS items. These production forecasts are added to the customer order requirements to arrive at a net demand statement. This approach is illustrated in Figure 4.44. In the example, a forecast is developed for the PC product family. Each sold PC is registered in the row 'Customer orders'. A backward/forward forecast consumption rule (as illustrated in Figure 4.48) was used to derive net demand. The unconsumed forecast represents that fraction of the forecast that has not been consumed yet by incoming orders. It equals the net demand minus promised orders. This unconsumed forecast is then exploded using a planning bill to corresponding production forecasts for the individual MPS items. The example assumes a planning bill with a 50% ratio of 386 PCs. The net demand for the '386 hardware unit' MPS item is then calculated as the sum of actual customer order requirements and the production forecast.

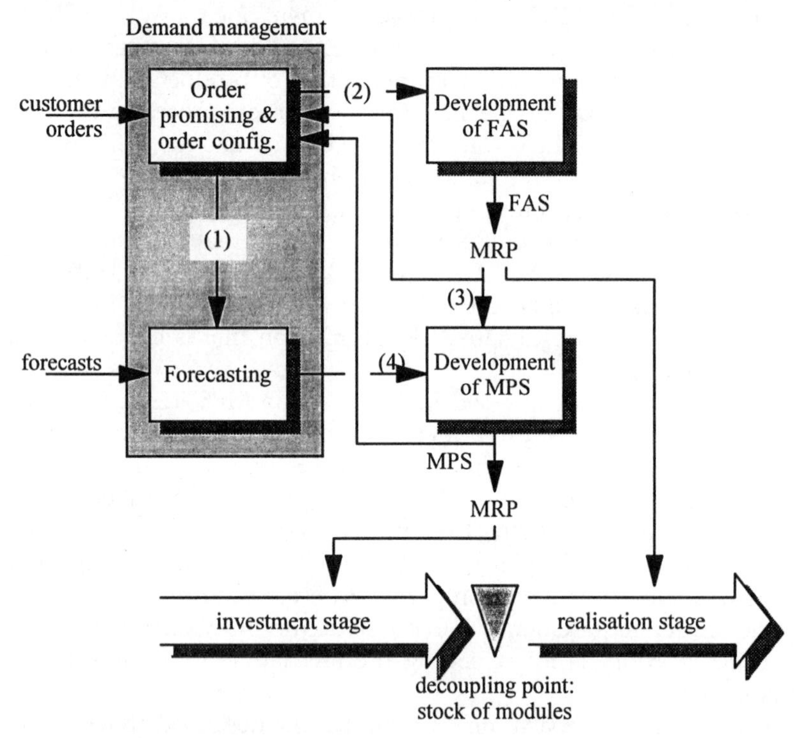

Figure 4.43 Master planning for an ATO environment.

PC product family (backward/forward forecast consumption)

	1	2	3	4	5
Forecast	10	10	10	10	10
Customer orders	8	20	4	4	
Net demand	8	20	4	8	10
Unconsumed forecast	0	0	0	4	10

50 %

386 hardware unit

	1	2	3	4	5
Production forecast	0	0	0	2	5
Customer order req.	4	14	3	1	0
Net demand	4	14	3	3	5

Figure 4.44 Net demand calculation for an MPS item in the ATO environment.

If planning bills can be used with relatively stable percentage parameters, the latter approach could be used. Forecast consumption is only done at the level of the product families and not for each individual MPS item. The production forecasts and the derived net demand statements for the individual MPS items all refer to one and the same basis. Forecast consumption at the level of the product families is especially relevant when the entire product range within the product family is equally relevant for the market being considered (e.g. cars with or without certain options).

However, if within a certain product family certain products are restricted to certain markets, forecast consumption at the level of MPS items may be considered. In the case of cars with left-hand drive and right-hand drive, the forecast for the left-hand drive option should be consumed independently from that for the right-hand drive option.

Note that the statements of net demand for the MPS items should take account of the final assembly lead time, before they can be used as an input to the MPS development function.

Order promising is done at the MPS item level. Incoming customer orders are configured in terms of MPS items requirements, which need to be compared with the ATP information. Figure 4.45 illustrates how this information is arrived at.

For purposes of simplicity, forecast consumption is done at the MPS item level. The resulting net demand must be offset by the final assembly lead time before it can be used as an input for development of the MPS. The net demand, after offset by the final assembly lead time, together with the dependent requirements in support of the FAS, constitute the total requirements that are to be satisfied by the MPS. The firm requirements derive on the one hand from the final assembly schedules of the finished products (20 units in week 1) and on the other hand correspond with the firm customer orders that have arrived beyond the final assembly horizon (16 and 8 units in weeks 2 and 3, after offset by final assembly lead time). Available to promise information is calculated in the usual way. ATP for week 1 is for instance derived as [the on-hand availability of 10 units] + [the MPS quantity of 40 units in week 1] – [the 20 units firmly required in week 1] – [16 units firmly required in week 2 (there is no supply in week 2)] = 14 units. This ATP means that 14 units of the 386 hardware unit are available to promise as of week 1. The products for which they will be used can only be promised as of week 3, when taking account of the final assembly lead time of 2 weeks.

PC 386 tower frame	p. due	week 1	week 2	week 3
Customer orders		10	10	10
Final ass. schedule		8	10	10
Scheduled receipts		8	10	
Pl. order releases		10		
PAI	2	0	0	0

On hand = 2, final assembly lead time = 2 weeks

386 hardware unit	p. due	week 1	week 2	week 3	week 4	week 5	week 6
Forecast before offset					20	20	20
Customer order req. before offset					16	8	
Net demand, before offset					20	20	20
Total requirements		20 (1)	20	20	20		
MPS		40		40			
PAI	10	30	10	30			
Firm requirements		20	16	8			
Available-to-promise		14		32			

On hand = 10, production lead time = 2 weeks
(1) requirements do also include requirements emanating from Final Assembly Schedules for other finished products

Figure 4.45 Calculation of ATP in the ATO environment.

4.3.7.3 Master planning for MTS production of modular products

Modular production appears to be very powerful in ATO environments. Modular production has the primary advantage that standard modules are produced in a first stage of production, which subsequently are used to produce a wide variety of finished products. Allocation of a module to a certain product variant is postponed until very late in the production process. Flexibility of allocation of materials is hence maximized.

The proposed **planning approach** for ATO companies entirely reflects the modular **production approach**. An MPS drives the stage of module production, whereas the realization stage is driven by the realization schedule.

Modular production may also be of interest to MTS companies, since it increases the flexibility and therefore also the responsiveness of the production system. The adopted planning approach for MTS companies does not however reflect the inherent flexibility of modular production. MTS companies build to stock. All of the production operations are executed to anticipate expected demand. Master production scheduling is executed at the level of finished products, and buffer inventories are planned at the level of finished products.

MTS companies with modular production could take advantage of the planning of buffer inventories at the level of modules. These buffer inventories are not committed to a specific finished product and may help increase responsiveness to changes in demand. However, since master production scheduling is executed at the level of finished products, buffer inventories for product modules are not clearly visible and hidden 'at lower levels' within MRP data. As a result, those buffer inventories cannot be immediately checked upon when promising extra demands or faced with demand mix changes. An alternative planning approach is therefore required when MTS companies want to take advantage of modular production.

Assume that the PC production company with the superbill structure of Figure 4.15 decides to reduce its product offering, and to produce prepackaged products to stock. Master production scheduling would then be executed at the level of finished products with any buffer inventories planned at that level. Even with a similar production approach (first module production, then assembly of finished products), the planning approach would change drastically. Any surplus inventories for the individual PC modules would no longer be visible at the MPS level, since master production scheduling for an MTS company is typically performed at the level of finished products.

The purpose of this section is to prove that it is actually possible to combine an MTS planning approach with ATO planning concepts to achieve a flexibility within planning, similar to the existing production flexibility inherent in production (to stock) starting from standard modules.

The planning architecture is represented in Figure 4.46. All operations are executed to anticipate demand. There is no realization schedule. Distinction is made between an investment stage for production of the standard product modules and an investment stage for assembly of the finished products. The two stages are driven by two master production schedules. One is expressed at the level of the modules, the other one at the level of the finished products.

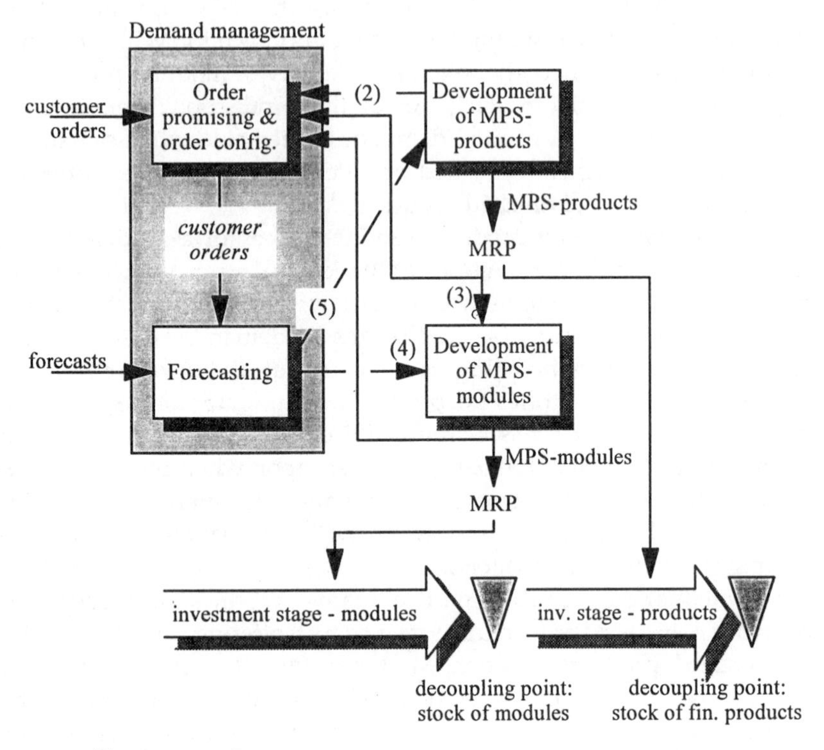

Figure 4.46 Master planning for the MTS environment with modular production.

The forecasting function develops the net demand statements for both the finished products and the product modules. The net demand schedule for the finished products is not developed across the entire planning horizon, but is limited to the assembly horizon. Net demand beyond the assembly horizon is expressed not at finished product level, but at the level of the modules. The net demands are input to development of the master production schedules for the finished products and the master production schedules for the product modules.

The master production schedules for the product modules have to satisfy not only the corresponding net demand schedules, but also the dependent requirements from the master production schedules for the products.

Order promising is based upon three inputs, i.e. the master production schedules for the products, the master production schedules for the product modules and the product module requirements, which are dependent upon the master production schedules for the products. In this way, ATP can be calculated for both finished products and modules. Depending upon **when** a customer order needs to be promised (either within or beyond the assembly horizon), order promising will be based upon the ATP information for the finished products or the ATP information for the product modules. The planning approach entirely reflects the flexibility of modular production. Safety stocks at module level are visible as ATP quantities.

The above concepts are illustrated by the example in Figure 4.47.

The first table illustrates how is arrived at the MPS for the PC386 tower frame product. The MPS is not expressed across the entire planning horizon, but is limited to the assembly horizon (in this example 2 weeks, with orders expected at the start of the relevant time bucket). The second table illustrates the development of the MPS for the 386 hardware unit. The net demand is one input to the MPS development process. Another input is the requirements that are dependent upon the master production schedules for the finished products. The third and fourth tables illustrate the calculation of ATP. The ATP calculation for the PC386 tower frame product is very similar to the ATP calculation within MTS environments. The ATP calculation for the 386 hardware unit follows the approach for ATO environments. The promised quantities are those to satisfy the master production schedules for the finished products (20 units in week 1, assuming that component requirements of the scheduled receipts have been issued from stores) and those corresponding with the customer orders promised beyond the assembly horizon (respectively 16 and 8 units). Again, it must be noted that module requirements need to be offset by the assembly lead time when promising incoming orders for finished products beyond the assembly horizon. A request for delivery of 6 units of the PC386 tower frame product could be satisfied starting from week 3 (5 units are promised out of ATP for week 3 and one unit out of ATP for week 1). A request for 20 units could be satisfied starting from week 5, assuming that all other modules are available and taking into account the assembly lead time of 2 weeks.

Development of FAS and MPS

PC 386 tower frame	p. due	week 1	week 2	week 3
Forecast		10	10	10
Customer orders		10	10	5
Net demand		10	10	10
MPS		12	10	10
Projected on hand		2	2	2
Scheduled receipts		12	10	
Pl. order releases		10		

On hand = 0, final assembly lead time = 2 weeks

386 hardware unit	p. due	week 1	week 2	week 3	week 4	week 5	week 6
Forecast before offset					20	20	20
Customer order req. before offset					16	8	
Net demand, before offset					20	20	20
Total requirements		20 (1)	20	20	20		
MPS		40		40			
Projected on hand	10	30	10	30			

On hand = 10, production lead time = 2 weeks
(1) requirements do also include dependent requirements from other master production schedules

ATP tables

PC 386 tower frame	p. due	week 1	week 2	week 3
MPS		12	10	10
Firm requirements		10	10	5
Available-to-promise		2		5

On hand = 0

386 hardware unit	p. due	week 1	week 2	week 3
MPS		40		40
Firm requirements		20	16	8
Available-to-promise		14		32

On hand = 10

Figure 4.47 Development of an ATP in an MTS environment with modular production.

The crossover between the horizon for the finished products and the horizon for the product modules needs to be managed carefully. Within the example, one could conclude that a maximum of (2 + 5 + 14) computers PC386 with tower frame could be promised within week 3: 2 and 5 units are still available to promise as part of the MPS for the finished product, and a further 14 units can be made available within an assembly lead time of 2 weeks (assuming that other product modules are available). The availability of these last units is, however, subject to the release of an assembly order of 14 units within week 1.

The planning approach has the major advantage that ATP of modules is visible as such. Within production, modules have not yet been allocated to specific product variants, and this is entirely reflected by the planning approach.

The presence of two separate types of master production schedules, i.e. the master production schedules for the finished products and those for the product modules, gives rise to some extra problems, in particular as regards the development of the master production schedules themselves.

The master production schedules for the finished products must be developed within the boundaries of availability of product modules. This is less relevant than in the ATO environment. Within the latter environment, customer orders are promised based upon ATP information for the product modules. If ATP information is adhered to, then product modules should normally be available to satisfy the production of promised customer orders. In other words, the feasibility of the realization schedule from a material supply point of view is guaranteed by the order promising process. In MTS environments with modular production, the situation is slightly different. Customer orders are also promised based upon ATP information, but the master production schedules for the finished products aim at more than only the realization of the promised customer orders. They also aim at satisfying a predefined responsiveness policy, which may be expressed by predefined PAI targets, and at meeting unexpected demands that have to be promised within the assembly lead horizon. Dropping PAI levels should be compensated by increased master production schedules, but these can only be increased within the boundaries of availability of product modules. In order to avoid a much too slow replenishment reaction, any shortfall of actual PAI levels versus PAI targets should be translated directly to extra requirements for the MPS modules. Any excess of projected product availability should on the contrary result in a downward revision of the master production schedules for the finished products, which would result in less requirements of product modules, which in turn would trigger a revision of the master production schedules of the product modules.

In this section, it has been shown that even within MTS environments a decoupling point can be made visible in the planning approach. Planning can in this way be much more intensively coupled with actual production, a necessary condition for fast and reliable order promising.

4.3.8 EVALUATION OF THE MRP II PLANNING APPROACH

4.3.8.1 Validity of the MRP II planning model

Master planning has so far been discussed within the MRP-based planning architecture, which is still considered as the favourite architecture for production planning and control. The logic of MRP is actually simple and straightforward, to such an extent that the core of MRP (the functionality of MRP I systems) could be referred to as a simple calculation logic.

The MRP model recognizes the concept of dependent demands: requirements of any component or subassembly item can be derived from the planned orders for the parent items in which it will be used. These planned orders aim at satisfying the corresponding requirements, which in turn were derived from planned orders for their parent items. As a matter of fact, one MPS is sufficient to determine all dependent requirements, if:

- at each level of the bill of material, planned orders can automatically be calculated to satisfy the corresponding requirements;
- lead times can be estimated by which planned order releases should be offset from their corresponding receipts.

These two assumptions prove to be the weaknesses of the MRP logic. Automatic calculation of planned orders requires that rules must be programmed as to how MRP should plan orders in response to requirements at a certain level of the bill of material. Orders are planned for each item individually using a cascade process from higher levels down to lower levels of the bills of material. The impact in terms of capacity requirements can never be assessed during the MRP calculation run. Lot sizing policies are part of the rules, and comprehensive evaluations of different policies have made clear that none of them is really optimal. Using a lot sizing policy is a simple but not entirely satisfactory method to, for instance, save set-up time for capacity constrained processes, or enforce minimum purchase quantities.

Lead times need also to be estimated. Actual lead times depend upon the level of resource utilization (these aspects have more comprehensively been covered within section 4.3.5.7). However, lead times need to be defined as fixed parameters, and it is quite logical to define these in such a way that even in situations of temporary overload actual lead times do not exceed planned lead times. MRP is essentially a logic to plan and control the flow of materials: component and subassembly production is planned to support the downstream operations. Excessive lead times within MRP harm the flow objective, since they result in work-in-progress (WIP) inventories between different operations.

The fact that MRP cannot work without WIP inventories and some 'overestimation' of actual lead times has clearly been demonstrated when discussing the process of developing a master production schedule. First, a materials plan is developed, which subsequently is evaluated for feasibility with respect to available productive capacity. MRP itself is capacity insensitive, and some internal slack is required in order to guarantee that bottleneck resources are continuously loaded, and that resource overloads can be smoothed over time.

The above limitations imply that the MRP logic is not suitable for every type of production environment. Companies with complex bills of materials and a project or job shop production layout often benefit very much from MRP II. These companies often have a large variety within their product offering. Primary focus is on the management of materials supply as well as the management of dependencies across the bills of material. The planned lead times within MRP also reflect the amount of queuing that is so typical for project and job shop environments. The top-down planning approach within MRP II fits well with the management philosophy of these companies. A materials plan is developed, which is roughly tested against availability of capacity. Dependent items are planned via MRP, and CRP is used to adjust capacity availability to calculated capacity requirements.

4.3.8.2 MRP II versus production by means of kanban

As indicated earlier MRP appears to be a good fit for companies with production planning primarily focusing on the management of materials. Those companies can be found within the segments of capital goods and durables.

However, it must be noted that MRP loses part of its importance within the durables segment. The companies within this segment are forced to produce a wide range of products, cost-effectively and with minimum customer lead times. The assumptions of MRP do not optimally support the kind of improvements required to safeguard competitiveness. MRP plans the flow of materials, while taking into account estimated lead times between the successive operation steps. These estimated lead times are chosen so that, even within periods of high load, actual lead times are within planned lead times, and production can be executed according to plan. This assumption typically results in inflated lead times, and corresponding high work-in-progress levels. It is true to say that substantial improvements in lead times and WIP levels have been made possible by the introduction of MRP. However, MRP appears to be an obstacle to further improvements.

Companies within the durables segment (e.g. hi-fi, camcorders, cars) normally produce at a stable rate of production for different product families. A product family typically consists of a range of products that are assembled from a set of standard modules. The Japanese were the first to introduce much simpler concepts than MRP to plan and control such environments. By introducing kanban, they actually returned to the much older reorder technique, however with buffer stocks no longer within a separate warehouse but physically between the successive operation steps. Buffer stocks are managed to hold a certain minimum stock level (the reorder point), and consumption of the stock to below this minimum level triggers production by feeding operations. The minimum levels for the buffer stocks between the successive operations, as well as the reorder quantities (as indicated by for instance kanban cards) are set as low as possible. The slack within the production environment is minimized in order to reduce work in progress and lead times.

MRP may partially or completely be eliminated with the kanban approach. The MRP explosion logic is – within the pure kanban environment – actually carried out by the kanban process itself, which aims at replenishing minimum work-in-progress inventories between the successive operation steps. However, with or without MRP, master planning keeps its dual role. Even more important than in other environments, the MPS is required to enforce stable production rates (mostly for certain critical production facilities, such as assembly tracks), which then are used as the basis for order promising. The production rates, in conjunction with explosion by MRP, may be used as the basis for negotiating supply contracts and calling off materials from suppliers.

The application of kanban techniques requires stable production environments with stable production rates. Production is managed in a repetitive mode. Minimum stock levels and reorder quantities for the different buffer stocks are set in accordance with the production rate. Highly variable production rates would require minimum stock levels and reorder quantities continuously to be reassessed.

With the introduction of JIT and corresponding kanban techniques, the production system is managed more from a capacity than from a materials point of view. This is clear when looking at the function of master planning. An MPS within a typical kanban environment serves to enforce stable production rates, often in response to pre-established sales agreements with important distribution channels. Order promising is much more about scheduling customer orders to fit with pre-established production programmes. The production programmes actually mirror the available productive capacity.

It is important to note that by adopting a kanban production approach, a production company positions itself between job shop and mass production in order to apply flow manufacturing principles to a 'wide' product range of 'similar' products.

4.3.8.3 MRP II versus synchronous manufacturing

The theory of synchronous manufacturing originated from the ideas of E. Goldratt and evolved in successive steps to what now has become a very interesting innovative look at production planning and control (Umble and Srikanth 1990). In the early steps of its development, synchronous manufacturing (theory of constraints, OPT) was perceived as a potential rival for the well-accepted MRP planning and control concepts. Synchronous manufacturing did challenge, and rightly so, some of the basic assumptions of MRP.

Synchronous manufacturing aims at the optimization of throughput, which is defined as 'the quantity of money generated by the firm through sales over a specified period of time'. Throughput is optimized by identifying the resources that could disrupt smooth product flow, and exploiting these as best as possible. The premise of the synchronous manufacturing approach is that only through scheduling all capacity constrained resources can be identified, and their exploitation be optimized. The approach therefore breaks with the traditional top-down MRP II approach but puts a finite capacity scheduling algorithm at the centre of the planning framework. The schedules for the capacity constrained resources are developed in such a way that they are robust with respect to unexpected changes in the production environment. They are used as a kind of drum beat for the other less or non-critical resources.

Finite scheduling within synchronous manufacturing differs from the more conventional Leitstand scheduling (which will be discussed in more detail in Chapter 5). The input for the synchronous manufacturing planning approach is **not** the set of MRP planned orders (as for finite capacity systems or Leitstand systems), but a set of actual or anticipated customer orders.

The position of synchronous manufacturing with respect to MRP and the strong disapproval of some of the basic MRP assumptions have clearly reinforced the impression of rivalry between MRP-based planning and synchronous manufacturing-based planning. Yet both have their own strengths and weaknesses, and as a result their own application areas. The core of MRP-based planning is the **material requirements** planning concept. MRP is in the first place focused at the management of materials in quite complex environments, characterized by complex bills of materials, many engineering changes, etc. The heart of synchronous manufacturing is a finite scheduling algorithm, which aims at optimizing the throughput in support of a set of actual and anticipated customer orders. Because of increasing uncertainty the more out into the future, the focus of synchronous manufacturing must be on the nearby horizon. As a result, synchronous manufacturing-based planning is a much better fit for companies faced with capacity constraints and involved in planning on a much shorter horizon than for MRP-based companies.

Important is that, with the introduction of the synchronous-based planning approach, a first step has been made in search of a new planning framework for companies for which the MRP II premises cannot be accepted. This issue is further developed in section 4.4.

4.3.9 FROM PRODUCTION PLANNING TO ENTERPRISE PLANNING

Within a project production environment, manufacturing involves more than only the production activity. Typical activities other than production are custom design or configuration, testing, installation and distribution. Master planning should therefore not be concerned only with the synchronization of production with demand. On the contrary, it must take account of all activities and manage the whole in response to actual and anticipated demands. These companies are exposed to the challenge of planning the entire enterprise, rather than only production.

The company producing digital exchanges that was introduced before is a typical and interesting project environment. Each customer order is a project in itself, with separate activities, which need to be executed in a predefined sequence. Some activities are executed as 'investment activities' (i.e. in anticipation of order confirmation), e.g. preliminary order configuration, module manufacture. Others are executed as 'realization activities' (i.e. after order confirmation), e.g. final assembly, testing and installation.

4.3.9.1 Master scheduling versus master production scheduling

Master scheduling for project environments is typically a hierarchical activity. It consists – as illustrated in Figure 4.48 – of a kind of network planning for each of the individual firm or anticipated customer order 'projects', with separate master 'configuration', master 'production', master 'installation' and other master 'activity' schedules at a second level. The master schedules at the second level need, of course, to support the overall master schedule at the first level.

The overall master schedule consists of the set of project plans for each of the firm and anticipated customer orders. Each project plan identifies the activities to be executed, indicates start and finish times for each activity, and shows their precedence relationships. Progress for each customer order is monitored against its project plan.

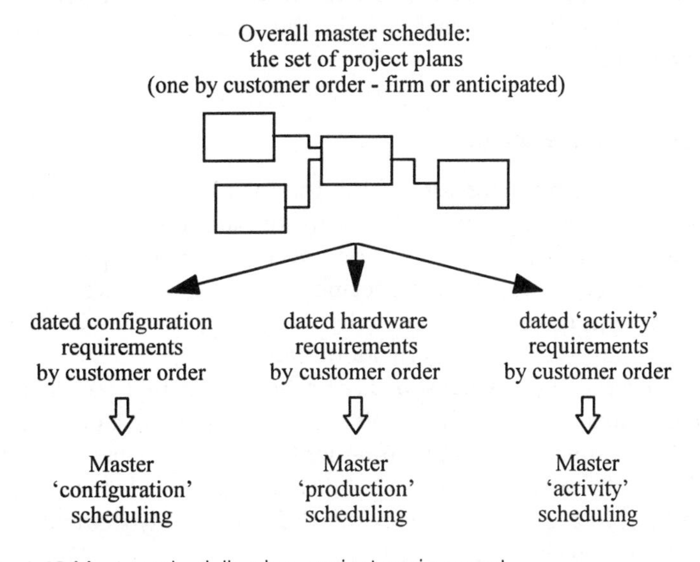

Figure 4.48 Master scheduling in a project environment.

By summarizing the activities across all of the project plans, the resource requirements of the overall master schedule on the individual departments (responsible for configuration, production, installation and other activities) can be derived. It is these demands that need to be satisfied by the development of achievable master 'activity' schedules.

The master **production** schedule at the digital exchange company introduced before is expressed in terms of the individual product modules. The MPS should satisfy the hardware requirements of all firm and anticipated customer orders. The hardware requirements are derived by considering the firm or preliminary order configuration for each firm or anticipated customer order, as well as the timing of the production activity within the corresponding project plan.

The configuration requirements could be a statement of required engineering hours by week. These again can be derived by summing the required engineering effort by project plan. The master **configuration** schedule could be as simple as a statement of number of engineering hours available per week, which ideally should be sufficient to meet the configuration requirements.

Generally, the requirements to be satisfied by the individual master activity schedules derive from the project plans, which together constitute the overall master schedule. The requirements include both firm and anticipated requirements corresponding with firm and anticipated customer orders. Order promising should ensure that customer orders are promised within the boundaries of each master activity schedule.

4.3.9.2 Demand management

Demand management consists of the forecasting and order promising functions. Forecasting is more than only the identification of expected demands. It includes the development of a project plan and the development of a preliminary specification by forecasted demand to ensure that requirements that are to be satisfied by the individual master activity schedules can be properly identified.

Order promising involves a verification of actual availability of resources (capacity, materials) against requirements as can be derived from the customer order. All requirements need to be promised against the individual master activity schedules. This is true for any demand that is added to the overall master schedule, or for any demand subject to a configuration change.

Evaluation of requirements versus availability of resources should be done based upon ATP logic for each of the individual master activity schedules. The ATP of the MPS would indicate the number of product modules that are yet available to be promised. Likewise, the ATP of the master engineering schedule could for instance indicate the number of engineering hours yet available to be allocated to a customer order.

4.3.9.3 MRP again questioned

Some articles advocate the use of the existing MRP system to model activities other than just production activities. All activities are structured in a 'bill of material'-like way and offset from the ultimate due date of the customer order. The authors are very sceptical about this approach not because of 'technical' unfeasibility, but because of the lack of control such a solution would provide.

It is true that a project network structure can be converted without too many problems into a 'bill of material'-like structure. Not all of the precedence relationships of project plans can be properly modelled using parent/child bill of material relationships, but all activity due dates can certainly be defined with a certain offset from the overall due date for the customer order project. A major shortcoming however is that these offsets are fixed parameters that cannot be adjusted to balance the load on the individual departments. A project planning level is mandatory in order to have the ability to stretch and/or shrink project plans to accommodate the limited resource availability within each department.

When working with fixed lead times between the different activities, everything is driven from one single customer order due date. All customer orders together may create a very unbalanced load on some or all of the departments. The authors believe much more in a first level of project planning and a subsequent level of master activity scheduling. Each customer order, firm or anticipated, has a project plan from which the requirements for the master activity schedules are determined. Promising should be done against the ATP information for each of the individual master activity schedules. Insufficient ATP may require that certain activities be planned earlier or later than originally intended and that the default offsets between activities be changed.

The problem with MRP is actually the same as encountered before. MRP in essence is capacity insensitive and works well when the primary constraint is not with capacity. Within project environments, a first levelling exercise to balance load across the individual departments is often a crucial issue.

4.4 IN SEARCH OF AN ALTERNATIVE PLANNING APPROACH

4.4.1 PLANNING REQUIREMENTS FOR PROCESS AND SEMIPROCESS INDUSTRIES

The discussion around MRP II has made clear that the MRP II-based planning and control framework is far from ideal across all industry segments. The core niche for the MRP II market is clearly within the project and job shop environments characterized by complex bills of material.

MRP's premises make it unsuited for most of the process and semiprocess industries. Process and semiprocess industries share a number of characteristics that require that the planning approach takes account of the specific nature of the production process. This is **not** the case with MRP II and is actually one of its major weaknesses.

The actual model used to drive the MRP engine is the set of BoM relationships, identifying parent and child items, combined with fixed lead times to offset planned start dates from planned receipt dates. The planned lead times are chosen so that a sufficiently large window is created to enable the timely execution of the production orders and to accommodate the queuing times before each of the work centres. The queues decouple in a certain way the consecutive operations on the shop floor. Determination of how the queuing orders are sequenced at each of the work centres is only done in a second step. Leitstand systems using rather simple sequencing rules may be used for this. This second step is a kind of suboptimization executed for the planned orders created by MRP in order to try to improve the throughput of released orders through the factory.

This two-step approach is mostly unacceptable for process and semiprocess industries. In contrast with many discrete production companies, (semi)process industries often tend to be very capital intensive, resulting in low amounts of excess capacity. (Semi)process industries typically produce a medium to wide range of similar products from a rather limited number of raw materials. The raw materials are often not so expensive and procured in bulk. Expensive or perishable ingredients can often be called off against pre-established supplier contracts. The management of raw materials is therefore often less of an issue with these industries and should not be taken as a starting point for planning purposes.

Other characteristics with (semi)process industries include limited shelf-life of semifinished products (expiration), sequence-dependent set-ups, recipes, etc. But one of the most important characteristics from a planning point of view is the requirement to adhere strictly to the production process, particularly in terms of routeings and operating instructions as well as min–max boundaries for actual operation and inter-operation times. In contrast with discrete job shop environments, (semi)process industries are characterized by batch or flow production. Planning needs to reflect the processing constraints. It is therefore not acceptable to start with developing a materials plan, based upon planned lead times about 20 times the size of actual operation times (as in a typical MRP environment), and finish by scheduling the MRP planned orders within the windows as defined by the planned lead times. Even if MRP was used with planned lead times nearly equal the actual operation times, no satisfactory solution would be obtained, since

MRP's logic is inherently capacity insensitive, an unacceptable condition when developing a plan within a (semi)process environment.

(Semi)process industries often have a second stage of discrete, often packaging activities. Such a second stage is more similar to the target domain of the MRP planning approach. The process stage(s) and discrete stage can sometimes be decoupled and planned independently from each other. In many cases, however, the output of the process stages needs to be packaged as quickly as possible, so that production planning needs to ensure the synchronization between the process stage(s) and the subsequent discrete stage. Again, MRP II planning concepts do not comply.

It may be clear from the above that an alternative planning approach is needed.

4.4.2 A PROPOSED PLANNING FRAMEWORK

Just as within discrete industry, planning for (semi)process industry needs to be done at different levels of detail and for different time horizons. Just as for discrete industry, planning needs to support the strategic, tactical and operational decision-making processes (Figure 4.49).

- Strategic: How are the individual business segments expected to grow? Can we cope with any increased demands? Do we need extra capital investments?
- Tactical: Which work pattern is needed across a medium-term horizon to cope with expected demands?
- Operational: What needs to be produced now?

These three decision-making layers do not accidentally correspond with the business planning, master planning and execution planning activities.

Business planning is required to support strategic decision making, which is concerned with shaping the business to increase competitiveness. Because of high capital intensity, the planning horizon may cover up to 10 years.

Master planning is needed to determine how to exploit available capacity using various combinations of work patterns to best respond to expected demands. Just as for discrete environments, master planning would also be used as the prime layer for order promising. The master planning horizon should extend beyond the cumulative production lead time and may cover up to 3 months. The master planning horizon within (semi)process industry is typically shorter than those in discrete industry owing to much shorter cumulative lead times. However, the horizon may be extended once per year to cover a full year to support the annual budgeting exercise.

Execution planning in (semi)process industry is – in contrast with execution planning in discrete industry – probably the most critical element in the production planning architecture. Finite resource scheduling should form the centre piece of the execution planning logic, so that finite capacity and rules with respect to set-ups, operation times, inter-operation times, etc., can optimally be taken into account. Leitstand techniques are *not* the solution, since their task is limited to considering the planned orders as created by MRP and sequencing the involved operations to optimize throughput. This suboptimal approach is not acceptable. The process requirements need to be taken into account from the start. It is not the planned orders as created by MRP, but the set of promised customer orders in combination with any extra expected demands, that should be the input for the execution planning exercise. The planning horizon should cover the cumulative production lead time.

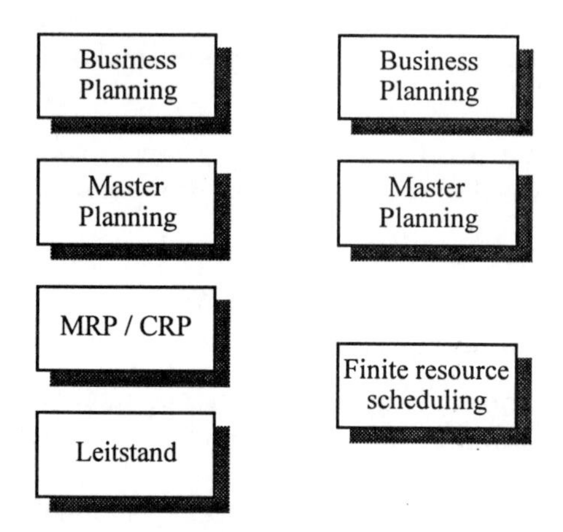

Figure 4.49 Planning layers within MRP II versus proposed planning framework for (semi)process industry.

As is indicated in the above text, execution plans may be developed to respond to a set of firm customer orders and a set of expected demands. In case of such a combination the company actually expects to receive extra demands over the near horizon to consume the forecast. It is clear that such a situation would only be accepted if to-stock production is allowed and finished products are not particularly subject to expiration. However, it also implies that part of the order promising (i.e. over the horizon as defined by the cumulative production lead time) may be done directly against the execution schedule. Order promising beyond the execution planning horizon would be done against the MPS.

If the MPS is to be reliable as a basis for order promising, it clearly needs to be developed while taking account of constrained capacity availability. Depending upon the accuracy with which promises need to be made (by day versus by week), it may also be necessary to take account of specific process requirements, which by the way are taken into account for purposes of execution scheduling. At first sight it would be ideal to apply the same planning technique for development of the MPS as the one used for development of the execution schedule. This way of working would be a good guarantee that the nature of the process is equally well taken into account as it is during development of a detailed execution schedule.

However, two things should be clear. First, an MPS does not serve the same purposes as an execution schedule. An MPS is not developed to determine when exactly production quantities will be produced and in what sequence. It is in the first place a statement of what will be produced by time buckets, which ideally are sufficiently narrow to be able to promise with sufficient accuracy beyond the execution planning horizon. Secondly, an MPS is developed over a much longer horizon than that of the execution schedule. This horizon is characterized by a lot more uncertainty. It would be wasted effort to develop a plan that is equally detailed as an execution plan.

Yet certain process requirements may still have to be taken into account, so that the traditional development approach within discrete industry would not be valid. Effort should then be undertaken to reduce the complexity of the execution planning model to obtain an MPS planning model that can be used with sufficient accuracy over a medium-term horizon for purposes of budgeting, shift planning and order promising.

(Semi)process industries are so diverse that no all-applicable guidelines can be formulated as to how to distil a proper MPS development environment from an execution planning functionality. Different scenarios exist, which all rely upon the trick to reduce complexity by aggregation of data. Possible scenarios are to aggregate by time (continuous execution planning versus bucket-wise MPS), by routeing (detailed routeings for execution planning versus simplified aggregated routeings for MPS), by machines (specific machines for execution planning versus machine centres for MPS) and by products (all items and products considered for execution planning versus planning by product family for MPS). Each of these scenarios needs to be tested in each individual environment. The ultimate purpose is to obtain a feasible approach for development of a plan over a medium-term horizon that is accurate enough for the purposes it should serve.

An important challenge within the (semi)process industry is to integrate properly the master planning and execution planning layers. Within the discrete industry, integration was quite simple. An MPS was developed, which served to drive MRP. The planned orders created by MRP served as the input for a Leitstand system, if any. The role of MRP (MRP I) in a (semi)process environment is – in the opinion of the authors – limited to the calculation of estimated requirements for purchased materials starting from the developed MPS. The MRP outputs are not used as the input for execution scheduling. An execution schedule is developed 'from scratch' by using finite capacity scheduling logic based upon promised customer orders and extra expected demands over a horizon determined by the cumulative production lead time. It is this execution schedule that could be used to call off materials against pre-established contracts.

The complete MRP II framework may however serve other important roles within the (semi)process industry. The MRP logic has been gradually extended into a comprehensive system of planning and control functionality, commonly referred to as MRP II. This functionality mostly includes – among others – cost control for production orders and provides for integration between the production planning and control functionality on the one hand and financial and cost accounting functionality on the other. Finite capacity scheduling systems do only focus upon planning functionality, and do not provide for extra logic, which may also be essential. The challenge resides in the proper combination of the planning functionality of a finite capacity scheduling system with the control functionality of MRP II. This asks for proper integration between both systems. A possible way of working would be to ensure that all created orders within the execution schedule are exported to MRP as firm planned orders or production orders before actually running MRP. This would ensure that, over the execution planning horizon, MRP would recognize the same orders as the execution schedule and would plan extra orders beyond. Within MRP II, all data would be available to allow tracking of production orders and integration with inventory management, financial and cost accounting functionality.

4.4.3 EXECUTION PLANNING TECHNIQUES

One of the most appealing areas for current research within the domain of production planning and control is focusing on appropriate execution scheduling techniques for (semi)process industries. It is a difficult quest, because of the variety of different (semi)process environments as well as the need to include their characteristics within the planning logic itself. The range of (semi)process industries is much more heterogeneous than the quite common job shop environment that is at the basis of the MRP II planning approach. Current developments in the area of finite resource schedulers have resulted in a number of different techniques that may prove more or less optimal for a given production process.

Simulation- and heuristics-based schedulers are often suited for processes that consist of different stages which by means of intermediate inventories can more or less be decoupled from each other. The process stages are scheduled the one after the other, either downstream or upstream of the production process. Depending upon the nature of each individual stage, scheduling may either be material dominated or 'processor' dominated. With a material-dominated scheduling approach, a materials plan will first be developed, which is subsequently tested with respect to available capacity. The opposite logic is used with processor-dominated scheduling, i.e. a plan is developed within the boundaries of available capacity, which subsequently is tested for availability of raw materials. Material-dominated scheduling is quite realistic for packaging lines. Instead, processor dominated scheduling is to be preferred for the typical 'capacity-constrained' process stages within a (semi)process industry. A plan for a stage determines the boundary conditions for the upstream or downstream stage, for which planning subsequently can be executed.

Simulation is used to search iteratively for a schedule that is feasible in terms of material and capacity requirements within each of the successive production stages as well as in terms of synchronization between them. The search may be facilitated by heuristics or algorithms translating, for instance, pragmatic rules.

Heuristics often use the 'drum–buffer–rope' logic introduced by the theory of synchronous manufacturing. The technique aims at the identification of the top bottleneck resource, for which a 'drum' schedule is developed. The drum schedule is used as the basis for development of schedules for feeding resources. The symbolism of a rope is used to represent the synchronization between the schedules for feeding resources and bottleneck resource. Whereas backward scheduling applies in front of the bottleneck resource, forward scheduling is used downstream of the bottleneck resource. Buffers are used to protect bottleneck resources from a temporary shortfall in supply.

Knowledge-based schedulers using expert system technology with constraint satisfaction techniques are often used in capacity-constrained job shop environments. **Stochastic schedulers** are based on techniques borrowed from other disciplines, such as simulated annealing (based on the 'natural' optimization that occurs during the slow annealing of fluid materials), taboo search and development of near-optimal schedules via genetic combination logic (simulation of the 'survival of the fittest' principle observed in nature).

Mathematical programming techniques, such as the simplex method for linear programming and the branch and bound technique for mixed integer linear programming, are the only real optimizers within the range of different techniques. A major disadvantage is that the calculation process cannot be interrupted to have a near-optimal solution.

Detailed descriptions of the above and other techniques can be found in specialized literature. It is important to repeat that, with the current state of development and knowledge, each of the above techniques need to be evaluated with respect to a given (semi)process environment. There is no standard way forward.

4.5 CONCLUSIONS

This chapter has extensively discussed the role of master planning, in particular within the traditional MRP II planning framework. The discussion has not only identified possible improvements in terms of how to use the MRP II architecture. Perhaps more importantly, it has helped make the boundaries of MRP II become visible. These weaknesses make visible the 'limited' applicability of the MRP II planning approach. They indicate the need for alternative planning approaches in process and semiprocess industries (Figure 4.50).

One of the major conclusions of this chapter should therefore be that it is possible to map planning approaches to areas of applicability. One of the most useful segmentation grids is represented in Figure 4.49, distinguishing production environments according to job shop or flow type of production and discrete, semiprocess or process nature. MRP II is very well suited for discrete job shop environments. Production planning for repetitive and continuous production is usually a lot simpler, just because their environments appear to be a lot more stable. Repetitive MRP (including backflushing techniques) may be used at the discrete side of flow shop production. It is interesting to see how JIT production and kanban techniques for discrete industries may be positioned between flow shop and job shop production. JIT aims at flow

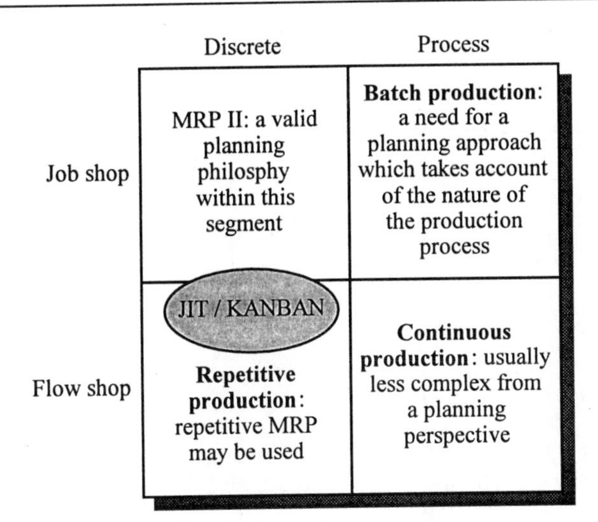

Figure 4.50 Segmentation of industries for evaluation of planning approach.

shop production with 'controlled' variety within a 'work cell'-based and/or job shop environment. Most planning complexity is with batch production. It is within the segment of batch production (segment of job shop-type production for process industry) that research needs to be focused to come up with alternative planning approaches. A short summary of currently available approaches was given in section 4.4, but a lot more work needs to be done. Finite resource scheduling techniques need to be improved even further, and knowledge and understanding must be increased so that the optimal technique and planning approach can be found more quickly and more reliably in accordance with the specific requirements of the production process.

Execution planning and control 5

5.1 INTRODUCTION

This chapter presents an overview of execution planning and control. The planning and control of shop floor activities has been the subject of much research and development in recent years. The area has been described extensively in earlier texts and numerous software systems and approaches have been developed. Some of this previous work will be used as examples throughout this chapter. The ideas presented in this chapter are mainly connected with the idea of integration with the higher planning levels, rather than presentation of new ideas within the domain of execution planning and control. However, some concepts may be viewed as being quite innovative in nature, and an attempt is also made to highlight any specific requirements necessary in different manufacturing environments mentioned earlier in this text.

The execution planning and control task resides at the lowest level of manufacturing planning and control and is often the area in which most problems are made visible. At this level machine breakdowns, operator unavailability, etc. occur, and these events cause disturbances in the system that affect the complete planning hierarchy. It is therefore essential that any plans downloaded to this level have enough flexibility to allow the operational planning and control function to operate effectively.

The requirements planning module in a manufacturing planning and control system typically performs the basics of a conventional MRP explosion, especially in the case of discrete parts manufacturing. This explosion process usually involves the explosion of MPS items into requirements for components and subassemblies. The explosion process may also involve the development of capacity profiles. The function of the execution planning and control system is to manage this exploded

MPS throughout the cells within a factory and perhaps between numerous factories. The transition from requirements planning to execution planning and control marks the transition from tactical planning to short-term planning and control. The problem is to ensure that the MPS that was exploded at the requirements planning stage is realized across the various work cells at the operational level of the factory. The concepts of factory coordination and production activity control have been well documented by Bauer *et al.* (1991). In their terminology the factory coordination system within the MPC hierarchy acts as a bridge between the requirements planning stage typically carried out by an MRP system and operational planning and control carried out by the production activity control system.

Factory coordination is described as a set of procedures concerned with planning and controlling the flow of products between different production units and which should have close links with the manufacturing systems design task. This design task is concerned with the design of the production environment in terms of the identification and maintenance of product families and an associated product-based layout. The complexity of the factory coordination task can be greatly simplified if the production environment is designed efficiently (Bauer *et al.* 1991).

It is vitally important for production controllers to remember that the requirements planning system produces planned orders. These planned orders are developed based on predetermined lead times and are used for both planned manufacturing and purchasing orders. The planned manufacturing orders are converted into actual orders by the factory coordination system. Actual orders refer to a 'filtered' set of planned orders, which are defined as necessary for release into the shop floor for execution. These actual orders are developed based on a more intimate knowledge of shop floor capability and constraints.

This more intimate knowledge of the shop floor is necessary at this stage of planning and control. Classical MRP theory does not cater very well for execution planning and control, and in many MRP systems this area is almost completely overlooked. Traditional requirements planning approaches (e.g. MRP explosion) have two main flaws that make them unsuitable as tools for control of the shop floor (section 2.4).

- Requirements planning does not recognize that lead times are variable. Therefore the estimates used within a requirements planning system are inherently inaccurate. This inaccuracy is acceptable at the tactical planning level but unacceptable at the shop floor or operational control (i.e. factory coordination and production activity control) level. MRP is not designed to be a shop floor control system as orders are positioned somewhere between planned start and due dates.

- The design of the production environment is typically ignored by requirements planning systems. Traditionally, manufacturing systems design was separated from the control of the product flow. Effective process design helps to eliminate a variety of possible production problems and in many cases may ease the control tasks involved. It should be clear that factors such as the layout of the factory and other structural issues need to be taken into account.

Many practitioners in this area assume that most of the operational planning and control problems can be solved by using software modules incorporating mathematically obtuse scheduling algorithms. Software solutions are very useful in the scheduling problem, but perhaps not so necessary for the other issues, such as dispatching jobs, moving materials, etc. For example, in the JIT kanban approach cards are the monitoring system and dispatching system (as they trigger the production). In many FMS implementations the dispatching mechanism for jobs between machines may involve a forklift truck with a human operator receiving instructions from the supervisor.

The mission of the operational or execution planning and control task is to allow for the realization of the longer term plans created at the tactical and master planning levels. These plans may be dissaggregated at the requirements planning level and then sent to execution planning for detailed capacity scheduling. Most MPS systems will perform a rough-cut finite capacity scheduling, i.e. capacity scheduling at a very aggregated level. This refined MPS can be regarded as an action plan. This is then used by the requirements planning modules to place orders on the MPS plan. The output of the requirements planning stage then becomes the input for a detailed fine tuning. This fine tuning should have the effect of aligning all the requirements planning output within the boundaries of the finite capacity that is available. This may involve explosion for material and capacity at the same time. This layer is effectively a translation mechanism between aggregate and detailed planning layers. Many analysis tools or approaches may be used within this layer, as depicted in Figure 5.1.

This list of requirements planning approaches is not exhaustive and mainly describes some support tools for use at the requirements planning level or as overall controllers of the execution planning and control problem. In Figure 5.1, a series of principles from different manufacturing planning and control approaches (JIT, OPT, MRP) are depicted. Certain principles and 'ideals' (e.g. zero inventory goal of JIT) from these approaches may be used together in order to fine tune an MPS. However, in general, there are only two main options available.

- The use of MRP as the 'overall controller' of execution planning and control. In this case, CRP is typically done after an MRP explosion process.
- The integration of tactical and operational planning by the use of detailed finite resource scheduling, for example, OPT.

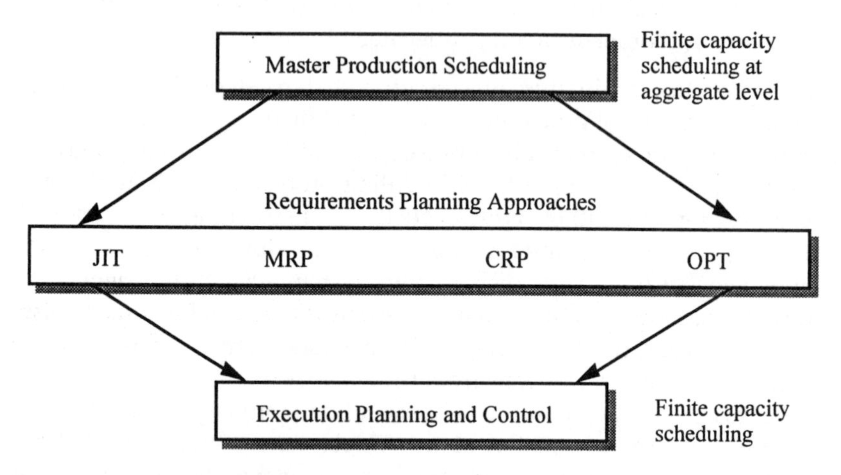

Figure 5.1 Requirements planning approaches.

The selection of the right approach to choose depends entirely on the manufacturing environment under consideration, and many 'blends' of MRP, JIT and OPT have been successfully implemented in practice.

In the remainder of this chapter, some techniques and approaches to the execution planning and control problem are described. The main approach that will be described here is the approach that was developed within the ESPRIT projects COSIMA ('Control Systems for Integrated Manufacturing' – No. 477) and IMPACS ('Integrated Manufacturing Planning and Control Systems' – No. 2338). This approach describes execution planning and control in terms of two levels, namely factory coordination and production activity control. We believe that this framework is quite generic in nature and we will use it as a reference framework throughout this chapter.

The remainder of the chapter is structured as follows.

- A framework for execution planning and control is presented together with an explanation of how this may include more than one level of planning functions.
- The impact of the design of the production environment on execution planning is explored.

- A well-known software support tool for execution planning is described, namely the Leitstand software.
- An example of an execution planning problem is explained. This example is linked to the detailed example presented in the previous chapter on master planning.
- Finally, the different requirements for small manufacturing enterprises (SMEs) are discussed.

5.2 A FRAMEWORK FOR EXECUTION PLANNING AND CONTROL

Within this section, a framework for execution planning and control within an MRP-like environment will be described in terms of factory coordination and production activity control. This discussion is based upon work done in the COSIMA and IMPACS projects. Firstly, a set of generic activities are described (releasing, scheduling, monitoring, moving and producing), which can be applied to any shop floor management and control organization. Secondly, the application of these activities may take place at more than one level of aggregation. This gives rise to the possible requirements for two levels, i.e. factory coordination and production activity control. Finally, the need for attention to the design of the production environment is also discussed.

5.2.1 A GENERIC FRAMEWORK AND SET OF ACTIVITIES

According to APICS, production activity control is a synonym for shop floor control, which is defined as:

> a system for utilizing data from the shop floor to maintain and communicate status information on shop orders (manufacturing orders) and work centres. Sub functions:

- assigning priority to each shop order;
- maintaining work-in-process quantity information;
- conveying shop order status information to the office;
- providing actual output data for capacity control purposes;
- etc.

The approach described by Bauer *et al.* (1991) presents execution planning and control as consisting of five major planning and control activities. These are:

- scheduling
- dispatching
- monitoring
- moving
- producing.

In summary, the **scheduling** activity is concerned with looking at events that will occur in the future, the **dispatching** activity attempts to control the present events, the **monitoring** activity attempts to provide analyses of the present and past activities, and the **moving** and **producing** activities help to implement the present events.

PAC has been defined as follows: 'PAC describes the principles and techniques used by management to plan in the short term, control and evaluate the production activities of the manufacturing organization' (Browne 1988). PAC resides at the lowest level of the architecture for PMS and therefore operates in a time horizon of perhaps 1 month to quasi-real time. It is desirable, for greater control, that PAC activities be as close to real time as possible, and consistent with actual manufacturing requirements. The PAC framework is illustrated in Figure 5.2.

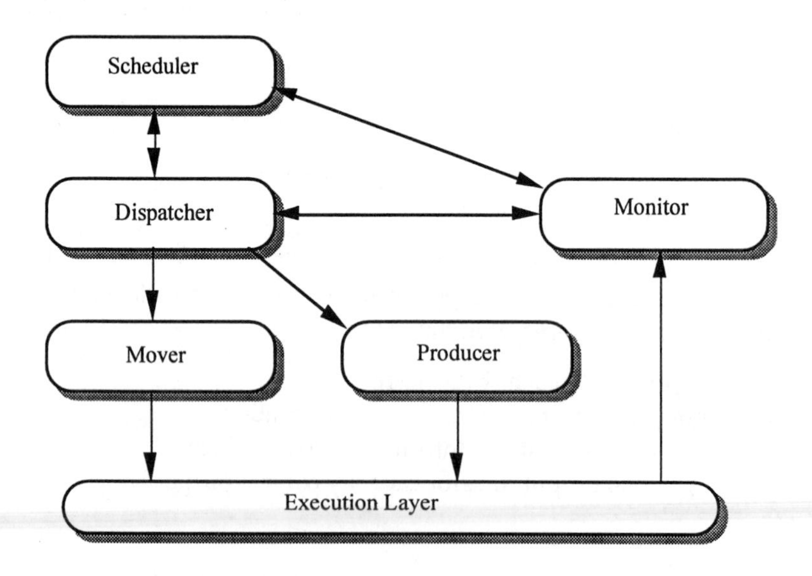

Figure 5.2 Production activity control framework.

Before the development of a schedule there may a need to process or filter the orders coming from the higher level. This may be done before the factory coordination or production activity control scheduling activity. This preprocessing is often described as a releasing function and this is suggested here as an addition to the framework.

- **Releasing.** The release function transforms the planned orders into actual manufacturing and purchasing orders. This is done by optimizing the allocation of available resources (materials, capacity) to planned orders and releasing only those for which resources are available or those that are part of urgent customer orders. It also adds factory-specific operational information to the planned orders information. All incoming planned orders are processed by the release activity. In effect, this function acts as a filtering mechanism before the task of detailed scheduling takes place.
- **Scheduling.** The scheduler takes the list of selected or 'filtered' order information and develops a detailed schedule for each work centre in the cell using finite scheduling techniques. This schedule is in fact a short-term plan that is passed to the dispatcher. The dispatcher can be considered to be a real-time scheduler that uses real-time status information from the shop floor – collected through the monitor – to dispatch individual batches to workstations. The dispatcher issues instructions to the movers and the producers on the individual workstations. In the case of factory coordination, the scheduler builds a factory-wide schedule of cell tasks and provides the user with tools for taking capacity and priority decisions at factory level. It also updates the schedule based on work-in-progress information. The scheduling function is activated every time new planned orders are released and every time the deviations from the previously established schedule caused by shop floor events require adaptations of this schedule. Planned orders are filtered into a set of actual manufacturing orders, for the factory's internal use, while the planned orders remain the reference point between the factory and the outside, especially the higher levels of planning and control. This activity generates also the cell orders to be scheduled and executed by the cell. At the PAC level, the schedule is developed based on the description of the processing activities of the product found in the manufacturing data, and a cell schedule is produced and sent to the dispatching activity.

Scheduling techniques often consists of two main categories: **backward** (or reverse) and **forward** scheduling. Backward scheduling often involves the use of Gantt charts (section 5.3). The development of these charts includes a technique by which the durations of past activities are subtracted from a required completion date. In forward scheduling the times are calculated by forward scheduling from a given date in order to obtain a completion date for a set of activities.

- **Dispatching**. The purpose of the dispatching function is to implement the schedule in the best way possible. It can be described as a technique through which it is possible to identify which of an available set of jobs to process next on an available resource. Typically, the objectives are, for example, to minimize throughput times, tardiness or the length of queues. At the factory coordination level, a dispatch list of cell orders is prepared for the PAC system controlling the cell. It maintains a continuous model of the factory global resources and integrates each new version of the factory schedule to its model. It then controls the workflow through the monitoring activity on an event-driven basis and triggers in advance the resources requirements in order to activate their transportation to the cell. This activity maintains lists of actions to perform and lists of events. At the PAC level, based on the status of the cells reported by the monitor, the dispatcher attempts to implement the cell schedule. It can take real-time decisions, such as choosing an alternative route, based on the process data found in the manufacturing data. It dispatches work instructions to the moving and producing activities and can request a new schedule from the scheduling activity when unexpected or 'out of control' situations arise.
- **Moving**. The purpose of the moving function is to plan and control the factory transport devices, i.e. the transport devices shared among several cells and transporting materials and tools between those cells. The mover receives requests for resources transportation from the dispatcher. These requests may consist of a plan of time-phased requests or single requests issued when required. This dynamic aspect depends on the moving activity planning capability, which is determined by the criticality of the transportation in the factory. The mover function implements the instructions from the dispatching activity. It controls the moving devices of the cell, playing a role of translator, and sends real-time reporting data to the monitoring activity, such as transporter device status messages.
- **Monitoring**. The monitor function is in charge of maintaining up-to-date information about all the entities considered by the system (cells, manufacturing orders and cell orders, time, etc.). This information may be obtained indirectly through reporting from the

PAC systems themselves. The monitoring function gives a picture of the current status of the factory. Furthermore, this information is stored in order to provide statistics on the past performance of the factory. Reporting uses the previous monitoring information, current or statistical, in order to provide various views of this information. The monitor receives various reports from the shop floor data acquisition systems and from the cells' PAC systems, prepares statistics and reports for factory coordination level and for upper levels. Monitored information is continuously maintained and reporting data is issued periodically or when required.

- **Producing.** The producing activity is comparable to the moving one. It implements the instructions from the dispatching activity. It controls the producing devices of the cell, playing a role of translator, and sends real-time reporting data to the monitoring activity, such as producer device status messages.

5.2.2 THE NEED FOR A MULTILAYER APPROACH

Within the framework of the IMPACS project, two levels were described within the execution planning and control system: factory coordination and PAC. In some cases, there is a need for coordination of the flow of parts and assemblies from machine to machine and also from cell to cell. In other words, factory coordination deals with the scheduling, dispatching, monitoring, moving and producing of entities between individual cells or shops and PAC deals in a similar fashion with entities existing within work cells (e.g. individual machines, operators, etc.).

PAC systems control production within each cell on the shop floor, but have no capability at a plant level. Factory coordination bridges this gap, and in general it provides a connection between tactical planning (in terms of infinite capacity, e.g. in the MRP II framework) and execution planning (in terms of finite capacity) by conversion of planned orders into actual orders, while taking into account the constrained availability of the required resources. The main themes of factory coordination proposed by Bauer *et al.* (1991) involve viewing the entire factory as one system (i.e. systems approach) and placing great emphasis on planning in order to reduce the complexity of the control task at lower levels. A link with a design module provides the capability of structuring the production environment through group technology principles. An architecture for factory coordination systems is illustrated in Figure 5.3 (adapted from Bauer *et al.* 1991).

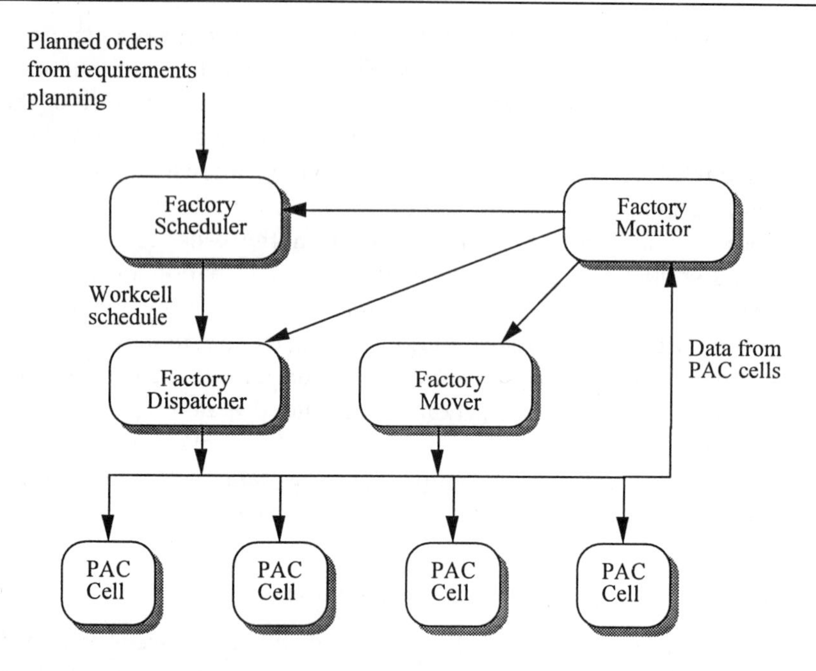

Figure 5.3 Factory coordination architecture.

As can be seen from Figure 5.3, the architecture contains four main elements, apart from the existence of a production environment design module. These are a factory-level:

- scheduler
- dispatcher
- monitor
- mover.

The factory scheduler function receives as input the planned orders from the requirements planning system (not necessarily an MRP system) on a periodical basis. In the case of many small manufacturing enterprises or certain process industries, for example, this could be a capacity planning or finite resource scheduling module that is specifically tailored for use in situations where materials are not the main constraints. This period may vary typically from 1 day to 1 month. These planned orders are often described in terms of required amount and due date only, but could include information concerning the associated customer orders. This would be very useful for tracking of customer orders throughout the execution cycle and in order to give a better picture of work in progress. However, if planned orders are received from an MRP system, in which lot sizing is used, then it is not possible to link planned orders with their individual customer orders. One planned order covers multiple requirements, which in turn are derived from other planned orders, which also cover multiple requirements and often include provisioning for replenishment of safety stocks.

The factory scheduler attempts to find the best possible solution to the scheduling and sequencing of these planned orders throughout the individual cells. The scheduler in fact creates a series of what are often called 'work orders' based upon whatever optimization criteria are currently requested. The optimization criteria may be very limited in the case of basic scheduling software modules, however very advanced mechanisms such as simulation software may be used to develop these schedules.

The work cell schedule developed by the factory scheduler is passed onto the factory dispatcher. In many real-life situations this may well be a human supervisor. In theory, the dispatcher is responsible for the real-time execution of the work orders received from the factory scheduler.

The factory-level mover is the module responsible for the movement of material between individual work cells. The dispatcher and mover functionality can be very similar in many cases. The factory monitor continuously creates reports and feeds back information to the scheduling, dispatching and moving functions.

In summary, the plan developed by the factory coordination system using the factory-level scheduler provides guidelines on the starting times for the jobs and batches in each work cell, so that the detailed schedule for each cell is developed in accordance with the overall strategy for the entire factory. The dispatcher passes the schedule down to the work cells while the monitor ensures that the scheduler and dispatcher are up to date with the status of the factory shop floor. Each individual work cell is then managed by its individual PAC system.

In so far as cells are independent, coordination is a relatively easy task. If the individual cells were independent, in that, for example, they did not share any resources and only completed items flow between them in well-defined patterns, then we could begin to look at the possibility of distributed requirements planning systems. The situation would then be analogous to that of a multiplant manufacturing firm that manufactures assemblies and major components at various locations and carries out the finished product assembly at a final assembly plant. In this situation each plant manages the BoMs for its 'products', and carries out requirements planning for these 'products' etc. This suggests that, in a situation of 'independent' cells, perhaps MRP would be run only at an aggregate level and that full BoM explosions (MRP runs) would be done at the individual cell level for each particular set of 'products' to be manufactured. This notion is illustrated in Figure 5.4.

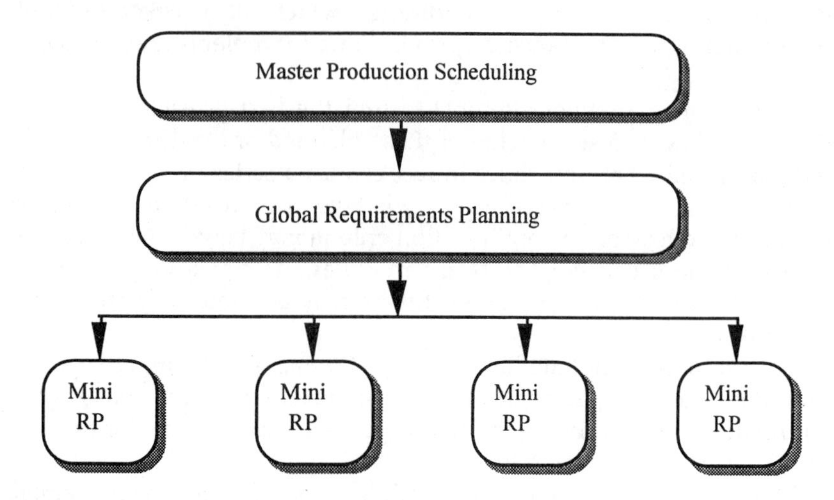

Figure 5.4 Distributed requirements planning.

In this figure, the master production scheduling function is seen providing the input to the 'global requirements planning' system. This system could effect a partial explosion down to a defined level in the overall BoM structure. It perhaps would not need to know all of the levels of components involved. This partial explosion would provide necessary aggregated data at that level. Then certain sets of the subassemblies or components would be identified as being the output of the individual cells. There then would be a detailed explosion process at each of these locations to enable effective planning and execution.

In general, the higher the degree of interaction between the manufacturing product-based cells, the greater the need for a more complex coordination system. Therefore requirements planning needs to perform its activities in a more detailed level, and to account for this added complexity there may be a need for a factory coordination system.

The overall factory-wide plan provided by the factory coordination system can be then downloaded to individual work cells. These work cells then have their own production activity control system to coordinate the flow of work within them.

5.2.3 IMPACT OF PRODUCTION ENVIRONMENT DESIGN

Effective design of the production environment may lead to a more easily controlled shop floor, and a regular and stable product flow. This can enable a more efficient introduction of new products and with less disruption to the existing process. The planning and control tasks concerning raw materials and work in progress become less complex. Owing to the stability of the production process, further production requirements for satisfying market demand can be determined with greater certainty. As demand fluctuates in batch manufacturing, the environment has the flexibility to handle the demand fluctuations because of the inherent stability and regularity.

A common approach to the problem starts with the identification of product families and flow-based manufacturing. The identification of product families and the subsequent development of flow-based manufacturing systems is often termed group technology (GT). The use of GT in JIT systems to define product families is important for a number of reasons. Firstly, GT is used to aid the design process and to reduce unnecessary variety and duplication in product design. Secondly, GT is used to define families of products and components that can be manufactured in well-defined manufacturing cells. The effect of these manufacturing cells is to reorient production systems away from the process-based layout and towards the product or flow-based layout. Group technology leads to cell-based manufacturing, which seeks to achieve shorter lead times, reduced work-in-progress and finished goods inventories, simplified production planning and control and increased job satisfaction. Group technology was not originally conceived by the Japanese, but its philosophy was adopted by them and drawn into the JIT approach to manufacturing.

It is useful to consider in some detail the differences between a traditional functional or process-based plant layout and a GT- or product-based layout. In the functional or process layout machines are organized into groups by function. Thus, in a metal cutting machine shop, the lathes would be grouped together in a single department, as would the milling machines, the grinders, the drilling machines, etc. A departmental supervisor or manager would be responsible for a particular function or group of functions. Individual components would visit some or maybe all departments and thus would pass through a number of different supervisors' areas of responsibility. Individual operators and their supervisors would be responsible for different operations on each component but not for the resulting component or assembly itself. Given the variety of components associated with batch-type production systems, the actual route individual batches take through the various departments or functions in the plant varies and therefore the material flow system is complex. Furthermore, given this complex and virtually random material flow system, it is not easy at any point in time to say what progress has been made on individual batches.

The product or cell-based layout is clearly considerably simpler than the process-based layout. In fact, this simplicity is a hallmark of just-in-time systems and for many writers and researchers on manufacturing systems, is a key characteristic of the system. As we shall see later this simplicity facilitates the use of a manual PAC system, namely kanban on the shop floor itself.

A technique used to plan the change from a process- to a product-based plant organization is production flow analysis (PFA) (Burbidge 1989). PFA is based on the analysis of component route cards which specify the manufacturing processes for each component and indeed the manufacturing work centres that individual components must visit. PFA, according to Burbidge, is a progressive technique, based on five subtechniques, namely:

- company flow analysis
- factory flow analysis
- group analysis
- line analysis
- tooling analysis.

Company flow analysis is used in multiplant companies to plan the simplest and most efficient inter-plant material flow system. Factory flow analysis is used to identify the subproducts within a factory around which product-based departments can be organized. Group analysis is used to divide the individual departments into groups of machines that deal with unique product families. Line analysis seeks to organize the individual machines within a line to reflect the flow of products between those machines. Tooling analysis looks at the individual machines in a cell or line and seeks to plan tooling so that groups of parts can be made with similar tooling set-up. Group technology in a sense creates the conditions necessary for JIT because it results in a better control of the variety seen by the manufacturing system, standardization of processing methods and better integration of processes.

In summary, the important issue from our point of view is that flow-based manufacturing is an important goal for manufacturing systems designers to aim at and it seems to provide added process control for execution planning and control purposes.

In this section, a generic architecture developed within the framework of previous work was used to describe the different modules contained in a execution planning and control system suitable for the larger manufacturing enterprises. Some of the issues relating to small manufacturing enterprises are discussed later in the chapter. The next section is devoted to a brief description of a currently popular planning methodology and potential solution for the execution layer.

5.3 LEITSTAND

One of the most outstanding tools that has emerged in recent years in the area of execution planing and control is the Leitstand. Leitstand is the German word for 'control post' or 'command centre' (Adelberger and Kanet 1991). These 'electronic control stations' (Köhler 1993) incorporate very user-friendly graphical displays based upon the classic Gantt chart format. The Leitstand can be described as the link between tactical planning and the control of the shop floor and includes the provision of manually editable schedules in graphical format. Leitstand-type systems are being used widely as a mechanism for distributed shop floor control in Germany, and many systems now offer a wide range of integration capabilities with standard MRP-type systems. One of the main reasons for the advent of the Leitstand approach was that highly automated scheduling systems typically focused on capacity balancing and sequencing and did not allow for random alterations involving human interventions.

Computer-supported Leitstand systems typically include five main components:

- a graphics component capable of representing the schedule in Gantt chart format;
- an interactive schedule editor that allows the user graphically to manipulate and change certain aspects of the schedule;
- a database management system that contains the information required by the scheduler;
- an evaluation component for performance measurement of schedules;
- an automation component for automatic generation of schedules.

The graphical component means that the user will not be confronted with technical skill requirements that differ greatly from those in use at present. It makes the scheduling process more intuitive for the user while also making it very easy for the scheduler to alter the schedule if required. When changing one job using the interactive scheduler, the user will see the effect of this change on all the subsequent jobs. An example of this graphical display is shown in Figure 5.5.

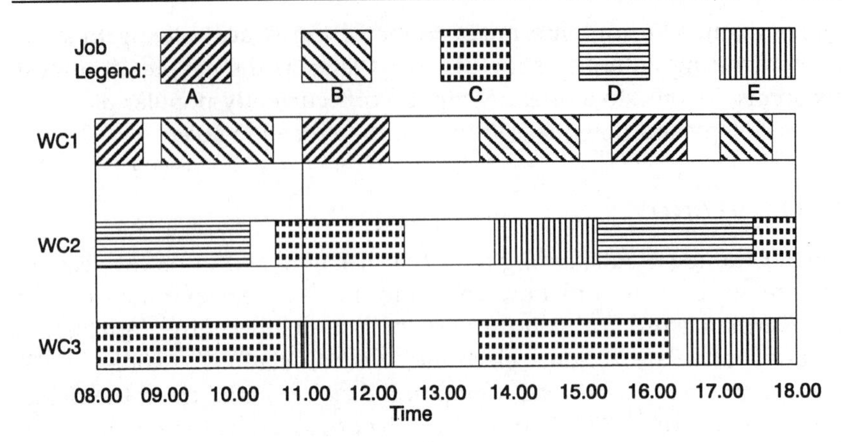

Figure 5.5 Basic display of a Leitstand-type system.

The contents of the display usually include a legend displaying a different colour or pattern corresponding to different planned orders to be scheduled. The vertical axis usually displays the machines available within the cell or workshop and the horizontal access displays the time of day. In our example in Figure 5.5, we have five jobs (A, B, C, D, and E) and three machining centres (WC1, WC2 and WC3). The systems are usually designed in such a way as to take into account the sequence of jobs across the different machines and set-up and transportation times may also be included. Maintenance schedules may also be incorporated into these systems. One of the obvious drawbacks associated with this type of systems is the amount of data that can shown on a screen at any one time. The Gantt chart representation is also used to display job history so that the human scheduler can very easily display the actual situation and compare this with the scheduled situation in order to assess performance. In effect, this could be classified as an extension to the capability of the monitor function with PAC.

The interactive scheduler is used to allow the knowledge of the human scheduler to be imposed on the scheduling process and to allow the schedule to change as disturbances affect the production system. This may be described as a 'simulation' capability of Leitstand-type systems. In effect, the systems could be described as discrete event simulators attempting to assign jobs to machines in the best possible sequence to maximize total throughput or to maximize the utilization of the resources. Interactive scheduling capabilities include the ability to manually 'drag' jobs from one machine to another using the mouse while keeping consistency with the routeing information.

The database management component deals with the storage of information relevant to the system in a consistent fashion. Performance evaluation and automatic generation of schedules provide extra capabilities for the users to review progress and to ease in the amount of manual intervention required. The scheduling procedure is typically heuristic driven and emphasis is placed on dialogue with the users rather than mathematically optimal algorithms.

In summary, the Leitstand or interactive scheduling approaches to execution planning and control offer many advantages, and most scheduling systems for any kind of manufacturing environment could incorporate some or all of the methods that are available. There are, however, some obvious drawbacks in terms of computer screen layouts and integration with higher level planning functions that need to be assessed before considering an implementation. In order to give a clearer understanding of the issues involved in execution planning and control an example is given in the following section which relates to the example given in Chapter 4.

5.4 EXECUTION PLANNING EXAMPLE

In Figure 5.6, the example bill of material and overall routeing information for product A, shown in detail in Chapter 4 (e.g. Figure 4.27), is repeated in order to give the context for an example of execution planning. In Chapter 4, this example was used to demonstrate the techniques of rough-cut capacity planning (at the level of the MPS) and capacity requirements planning (at the level of the materials requirements plan).

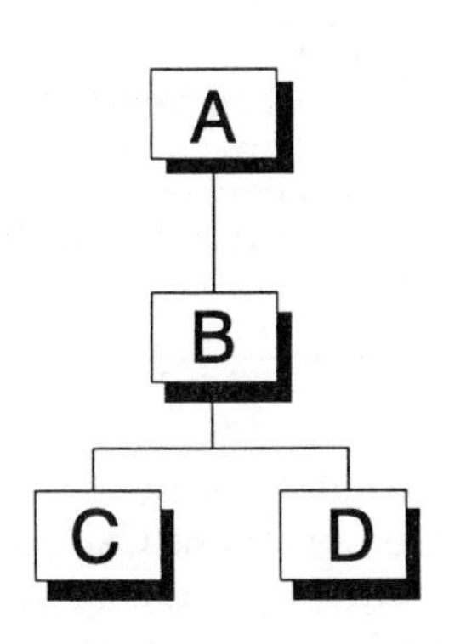

	Operation No.	Work Centre
A	1	WC1
B	1	WC1
	2	WC2
C	1	WC1
	2	WC2
D	1	WC2

Figure 5.6 Example bill of material and routeing information.

In the following example the numbers shown in brackets should be interpreted as follows: (1:x; 2:y) means that operation 1 is due at time x and operation 2 is due at time y. For example, in Tables 5.1 and 5.2, open orders and planned order releases as developed as part of the example presented in Chapter 4 (Figure 4.32) are summarized with this information shown in brackets.

If we look at the example of open order number 1003, we can calculate the due times of each operation as follows. Operation 2 is due at t=80 (the planned order receipt time; working weeks of 40 hours), whereas the due time of operation 1 is calculated as follows:

operation 1 is due at t=80 –
processing time for operation 2 (=10.5 hours) –
set-up time for operation 2 (= 8 hours) –
queue time for work centre 2 (= 30 hours)=
31.5 hours.

where processing time for operation 2 is obtained as follows:

$$\text{order quantity} \times \text{run hours per unit} = 210 \times 0.05 = 10.5$$

The work centre queue times for this example are 20 hours for work centre 1 and 30 hours for work centre 2.

The operation dues dates are used as the basis for the earliest operation due date sequencing rule that is used with this example.

Table 5.1 Example of open orders status

Order no.	Item	Quantity	Status
1000	A	75	Completed
1001	B	200	Operation 1 completed (2:40)
1002	C	180	Operation 2 past due and yet to be started (2:0)
1003	C	210	Not yet started (1:31.5; 2:80)
1004	D	180	Completed

A scheduled receipt is scheduled to be received at the start of the time bucket in which it appears. This means that the operations take place in the preceding time buckets. The scheduled receipt of 75 units of product A in time bucket 1 is scheduled to be received at the start of week 1. Therefore, the required operations should already have been completed.

This open order no longer creates any load on work centres. The open order of 200 units of B is scheduled to be received at the start of week 2. There is an assumption that operation 1 has been completed and operation 2 still has to be executed during week 1.

Table 5.2 Example of planned order releases status

	t=0		*t*=40		*t*=80	
Item	Order no.	Qty	Order no.	Qty	Order no.	Qty
A	1005 (1:40)	95	1008 (1:80)	85	1010 (1:120)	90
B	1006 (1:64; 2:120)	160			1011 (1:135; 2:200)	220
C			1009 (1:113; 2:160)	180		
D	1007 (1:80)	190			1012 (1:160)	190

A planned order release is planned to be released at the start of the time bucket. The planned order of 180 units of C in week 2 should be released at the start of week 2, with operation 1 being executed during week 2 if possible and operation 2 being executed subsequently.

The purpose of this example is to develop a schedule for execution of the open and planned orders. In Table 5.3, the sequence of operations by work centre are shown. This is also shown graphically in Figure 5.7. There are 40 hours per week available per work centre. The inventory evolution is depicted in Table 5.4 for items A to D inclusive. This inventory evolution is to be maintained during execution scheduling, since operations can only be executed if the required capacity and materials are available.

At $t = 0$, there are three **open** orders:

- order number 1001 of 200 units of B with operation 2 yet to be executed;
- order number 1002 of 180 units of C with operation 2 yet to be executed;
- order number 1003 of 210 units of C with both operations yet to be started.

There are three new orders **planned** for release, namely:

- order number 1005 of 95 units of A;
- order number 1006 of 160 units of B;
- order number 1007 of 190 units of D.

Table 5.3 Operations sequence and stock information

Item	Order No.– oper. no. (due time)	Qty	Start time	Material consumption B	Material consumption C	Material consumption D	Process time	Expected finish time	Destination
Work centre WC1									
C	1003-1 (31.5)	210	0				23	23	WC2
A	1005-1 (40)	95	23	−95			15.25	38.25	stock+95
B	1006-1 (64)	160	38.25		−160	−160	10	48.25	WC2
C	1009-1 (113)	180	48.25				20	68.25	WC2
A	1008-1 (80)	85	68.25	−85			13.75	82	Stock+85
A	1010-1 (120)	90	82	−90			14.5	96.5	Stock+90
B	1011-1 (135)	220	96.5		−220	−220	13	109.5	WC2
Work centre WC2									
C	1002-2 (0)	180	0				17	17	Stock+180
B	1001-2 (40)	200	17				32	49	Stock+200
C	1003-2 (80)	210	49				18.5	67.5	Stock+210
D	1007-1 (80)	190	67.5				25	92.5	Stock+190
B	1006-2 (120)	160	92.5				26	118.5	Stock+160
C	1009-2 (160)	180	118.5				17	135.5	Stock+180
D	1012-1 (160)	190	135.5				25	160.5	Stock+190
B	1011-2 (200)	etc.							

A dispatching rule must decide the sequence in which to take the orders in the case where there is more than one order waiting in a queue.

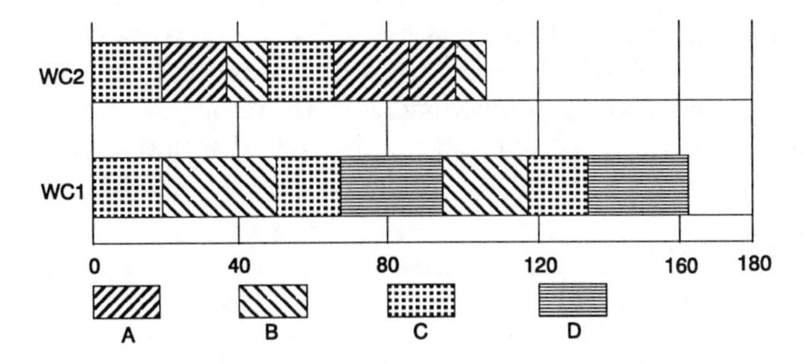

Figure 5.7 Gantt chart representation of scheduling example.

The 'earliest operation due date' is a commonly used and useful heuristic for this purpose. In the case that all operations are of equal importance with respect to due date, or any other criteria, an arbitrary decision is made.

Work centre 1 is first loaded by operation 1003-1 which is – with a due time of 31.5 hours – first due among the operations 1003-1, 1005-1 and 1006-1. This operation does not require any material consumption of other components and has a processing time of 23 hours. Work centre 1 is subsequently loaded by operation 1005-1 (due time = 40 hours) for a quantity of 95 units of A. The production of A requires 95 units of B, which at $t = 23$ are actually available (Table 5.4). The projected availability of B is expected to drop at $t = 23$ from 100 to 5 units. The total expected load on work centre 1 so far equals 38.25 hours, i.e. the result of the CRP run on open and planned orders for time bucket 1 (Figure 4.33). Work centre 2 is loaded in the same way as work centre 1. It is first loaded by operation 1002-2, then by operation 1001-2. The loads on work centre 2 can also be compared with the CRP results of Figure 4.33.

At $t = 40$, two new orders are released (Table 5.2). The orders are for 85 units of A and 180 units of C (order numbers 1008 and 1009). Operations 1008-1 and 1009-1 are therefore added to operation 1006-1 for scheduling on work centre 1, whereas 1009-2 is added to the list of queueing operations on work centre 2 (1003-2, 1007-1 and 1006-2). On work centre 1, first operation 1006-1 with a due time of 64 hours is scheduled. At $t = 48.25$, no sufficient materials of item B are available to process item A (at $t = 48.25$, the projected inventory for B is 5 units, whereas 85 units are needed to launch the order). The actual reason for the non-availability of B is the temporary overload on work centre 2. Work centre 2 has a backlog work of 17 hours (for operation 1002-2), with items B only becoming available at $t = 49$. The MPS has a requirement for 80 units of product A (Figure 4.32) at the start of week 3 ($t = 80$). These units can only be made available at $t = 82$. An alternative and better solution would be to stop production on work centre 1 from $t = 48.25$ to $t = 49$, when the B parts are expected to be received. Then production of A could immediately start, resolving the temporary shortage. Also, the delay for order number 1009-1 would not be harmful, since operation 1009-2 is only scheduled to start at $t = 118.5$.

Table 5.4 Inventory information

Item A			Item B		
Time	ΔPOH	POH	Time	ΔPOH	POH
0		10	0		100
0	+75	85	23	−95	5
0	−80	5	49	+200	205
38.25	+95	100	68.25	−85	120
40	−80	20	82	−90	30
80	−85	−65	118.25	+160	190
82	+85	20			
96.5	+90	110			
120	−90	20			
Item C			**Item D**		
Time	ΔPOH	POH	Time	ΔPOH	POH
0		0	0		10
17	+180	180	0	+180	190
38.25	−160	20	38.25	−160	30
67.5	+210	230	92.5	+190	220
96.5	−220	10	96.5	−220	0
135.5	+180	190	160.5	+190	190

POH, projected on hand

As can be seen from this brief example, even with such a small number of work centres, open orders and planned order releases, the scheduling task can be quite complex. Leitstand-type solutions can be very useful to represent graphically the problems that arise, therefore enabling a proactive approach to solving conflicts in the schedule. In some real-life situations where, for example, there are only few resources and a small product range, such as in the case of a small manufacturing enterprise, a detailed computer spreadsheet model may be sufficiently capable of providing the scheduler with a useful system for this task. In the following section, the focus is given to small manufacturing enterprises and the issues involved in execution planning and control for these types of companies.

5.5 EXECUTION PLANNING AND CONTROL IN SMALL MANUFACTURING ENTERPRISES (SMEs)

Often the production management strategies and solutions applicable to large manufacturers may not be suitable for smaller companies. In this section, developments on the adaptation of the factory coordination/PAC architecture to suit the needs of SMEs are presented. In many SMEs the human is the main source of expertise and knowledge within the PAC system. Within the framework of SMEs, it is vitally important that the human expert can exert control. When the freedom to control the system is given to the human expert it is important that sufficient decision support is provided so that well-informed and well-guided decisions may be made when planning and when reacting to unexpected events.

SMEs tend to exhibit some differences from larger companies (Jordan *et al.* 1993). Cross-functional management is not unusual and they tend to have single-cell production systems. Also, the products of SMEs tend to be less complex in structure than those produced by their larger counterparts.

The PAC architecture described earlier is particularly suitable for large, highly automated, manufacturing systems. There are many differences between these large manufacturing organizations and SMEs. Some of these are listed below. These differences are seen as typical but not absolute and are based on practical experience.

- Large organizations tend to be MRP driven in that they will tend to have an MPS detailing the aggregate production requirements for a long-term period and will use this to plan both short-term production and materials requirements. On the other hand, SMEs will tend to produce a less complex product by nature, often have a shallow BoM and tend to be customer order driven.
- Most SMEs can be described as being 'to order' in terms of the manufacturing environment.
- Large organizations tend to have a hierarchical organizational structure with a number of levels between top management and the operators on the shop floor. Smaller enterprises have a flatter organizational structure as they have fewer employees and as a result managers perform a broader range of activities than their counterparts in a large firm.
- Large organizations will tend to have a high degree of product variation and will thus often be multicellular in nature, while SMEs will more often contain a single cell and produce a narrower variety of products.

- Large organizations may have the need and the resources to be highly automated, whereas SMEs may not be able to justify the expense of automation because of their size and turnover.

To adapt the PAC architecture for SMEs we must take account of these differences. As SMEs tend to be order driven we cannot assume that MRP output is available to the PAC system. In this case the concept of an 'aggregate production plan' is often used. This aggregate production plan can be described as a forecast of likely production over some future period. A system based purely on the PAC architecture could allow nervousness in favour of achieving due dates. An attempt should be made to minimize this nervousness and balance production at the aggregate production planning level across the scheduling period, as this represents the manufacturing response to forecasts and/or customer orders. In order to achieve this, the concept of workload control can be used at the shop floor level in an attempt to stabilize the load by determining material and capacity requirements (Bertrand *et al.* 1990).

In SMEs, because of the lack of automation, the supervisors or managers tend to have far more control over production. In order to support this, shop floor personnel require relevant and accurate information. To provide this information, the manufacturing impact analyser (MIA), a shop floor decision support tool, was introduced to the PAC architecture. The architecture, as adapted to the SME environment, is shown in Figure 5.8.

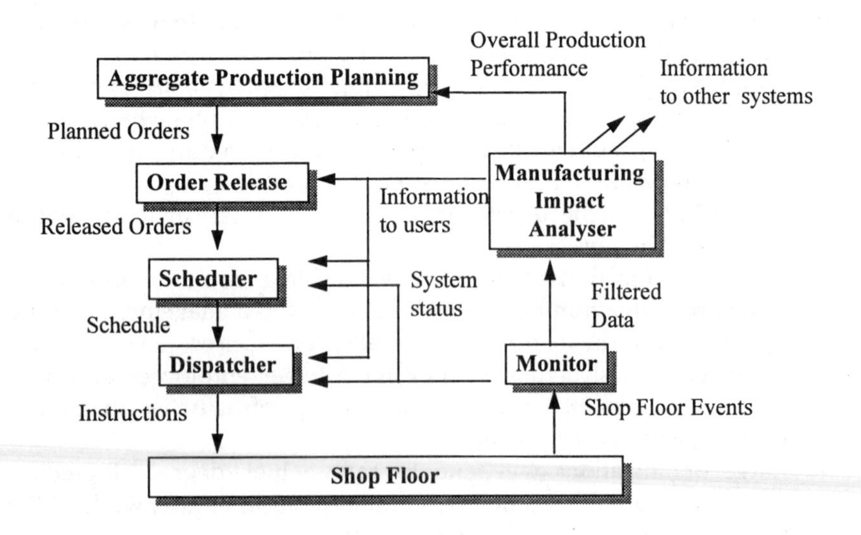

Figure 5.8 PAC architecture for SMEs.

The architecture still contains the essential elements of scheduler, dispatcher and monitor, but as SMEs are rarely highly automated the mover and producer elements were removed. These are replaced by the shop floor element in Figure 5.8. The main purpose of the MIA is to provide decision support to the human scheduler. To do this 'what if?' analysis tools and information filtering tools are provided. These include:

- a simulation tool that allows the use of trial and error techniques to predict the effect of changes or unexpected events in the production system;
- a backward scheduling tool for order verification;
- an information filtering capability that filters information according to the user of the system.

Different people within an SME require different information about the production system. The production supervisor may want to know where a particular order is within the system, while the managing director may only want to know the success rate at meeting delivery dates.

The MIA also exhibits the ability to source information; in effect, when information at a higher level is required, the user is allowed to reverse the filtering process and is guided down through the levels to the source of the higher level information. The concept of the MIA can of course be applied in any type of company, not just SMEs.

5.6 CONCLUSIONS

Leitstand systems and/or factory coordination/PAC solutions expect to receive planned orders as input from an MRP-like system. The role of the Leitsand/PAC system is to sequence these planned orders in a reasonable fashion and release them as manufacturing orders to the shop floor. In effect, this represents a kind of suboptimization that takes place after the MRP processor has exploded the requirements determined by the MPS. This approach is suitable for companies that operate in a material constrained environment.

On the other hand, finite capacity scheduling systems are important for companies in which there is mainly a capacity constraint problem. All activities are scheduled while at the same time taking into account the bill of material and routeing information. In this case, the explosion process takes place at the same time as the scheduling process. This approach is suitable for (semi)process industries. The MRP, Leitstand/PAC solution is not a good one for the (semi)process company and the factory coordination/PAC architecture is really only suitable for discrete manufacturing industry and these solutions appear as layers in a more complete MRP II architecture.

In our opinion, the framework for execution planning and control of factory coordination and production activity control presented in this chapter can be applied to any type of discrete manufacturing company. It is sufficiently generic to support the design and implementation of systems and is well documented in research work. The Leitstand systems offer extra features and a more user-friendly method of visualizing and manipulating schedules and this graphical aid can be of great use in the scheduling process. There are also other important issues that must be taken into account when dealing with this execution scheduling and control problems. These include the design of the production environment and the needs of smaller enterprises that may not have the need for large scale MRP II-type systems.

Overall conclusions 6

This chapter presents the main conclusions. A radical rethink is needed in the design and implementation of manufacturing planning and control systems in order to cater for the different manufacturing environments and the challenges faced by companies in today's competitive marketplace. The main themes developed in this book included a description of the changing manufacturing environments and their importance for manufacturing planning and control. The actual nature of the manufacturing environment should clearly be taken into account during the design (and implementation) of the planning and control system. Integration between the different levels (strategic, tactical and execution) also needs to be given high priority. As well as these overall themes, the book also was meant to demonstrate the limited applicability of certain approaches such as MRP II. Standard tools like MRP may be included in an overall manufacturing planning and control architecture that should firstly be tailored to the industry specific requirements. Finally, some new ideas were proposed to improve the planning activities at the strategic, tactical and execution levels in a manufacturing organization.

In this book, many new ideas were introduced and many old ideas were restructured to fit within an integrated architecture. The remainder of this final chapter contains an executive summary of the main conclusions arising from the discussions in previous chapters.

- In general, the classical MRP II approach has not catered well enough for the strategic and execution planning and control activities in many enterprises. The main argument behind this was that the basic ideas of MRP are only appropriate at the tactical planning level. The MPS needs to be recognized as the key node in a manufacturing planning and control system.

- Decisions taken at the manufacturing business planning level serve to drive or guide the effectiveness of the lower planning levels. The activity of manufacturing business planning formulates appropriate manufacturing policies, validates their likely effectiveness and develops plans for their implementation.
- Strategic decisions tend to be overly influenced by costing data. Strategic policies are not always formally addressed, nor are their implications properly understood by manufacturing management.
- Changing customer preferences, reduced payback periods and more costly investments increase the importance of good decision making. The quality of decisions made at the manufacturing business planning level can be greatly improved by the use of modelling. Modelling provides an excellent means for strategic decision makers to understand their own manufacturing environment and the likely implications their decisions will have.
- Business planning was described via the use of a four-task model for resource planning, cost control, performance measurement and policy making. The four-task model is presented as a simple planning methodology and application that overcomes the biases with existing management control systems and empowers the management team with the information and explanations to successfully accomplish their tasks. The performance measurement system monitors the effectiveness of the adopted strategic, tactical and operational policies. In addition, it serves to integrate decision making across the three layers. This is achieved through the use of an objectives network.
- The role of traditional product costing is questioned. The implications of undercosting, overcosting and cross-subsidization are discussed. An understanding of work and resource behaviour is proposed as a prerequisite to understanding cost implications.
- Decision making is differentiated from information management and six manufacturing business planning decisions are presented: make or buy, process choice, infrastructure, capacity, facilities and organization design. These categories provide a useful framework that can be used to identify the broad range of strategic policies that influences the manufacturing system. Many of these policies have tended to be evaluated on the basis of qualitative criteria only. The adoption of the proposed four-task model enables the management team to validate its selected policies on a quantitative basis.
- Master planning is given much attention in this book. It is at this level that most of the intelligent decision making is converted into quantitative information for planning and control purposes. The

MRP planning architecture is described as a valid approach for companies primarily focusing on the management of materials, but another approach is recommended for capacity-constrained companies. Companies working with MRP can improve their performance by careful selection of a decoupling point and structuring of bills of material. The primary instrument for controlling companies with MRP-based planning proves to be the MPS. Not only is the MPS the driver of MRP, but it should also be used as the basis for order promising. Also, forecasting needs to be executed at the right levels of aggregation so that it can provide the required information with a sufficient degree of accuracy for the development of a truly valid MPS.

- Master planning has two major roles as an interface between the production environment and its market. The first role is one of synchronization. Master planning should – through the MPS – ensure that production respond in a concerted and optimal way to firm and expected demands. The second role is one to support order promising. The MPS should be used as the prime basis to promise delivery dates for incoming customer orders.
- Business policies and not the production plan are the linking glue between business planning and master planning.
- The decoupling point determines the degree of interaction between production and the customer. The position of the decoupling point is a strategic decision and may be different from product to product. It determines the customer order lead time, as well as the level at which inventory levels will be held and the MPS will be expressed. In seeking an optimum combination of responsiveness and low cost, companies should have a high interest in positioning their decoupling points at the levels of 'lowest commonality' within their bills of material.
- Master planning in project environments can consist of a master scheduling and master production scheduling layer. The master scheduling layer aims at developing a coordinating plan across all activities (production and non-production) within the company. The MPS is the response of the production environment to the production requirements as can be derived from the master schedule. Any new firm or expected orders should be added to the master schedule, but this can only be done after satisfactory promising against, in particular, available-to-promise quantities of the MPS.
- The MRP II planning framework is useful for companies with a predominant materials management problem and complex bills of

material. An alternative planning framework is needed for (semi)process industries, which have a predominantly capacity management problem.

- The framework of factory coordination and production activity control presented in this book can be applied to any type of discrete manufacturing company. It is sufficiently generic to support the design and implementation of systems and is well documented in research work. The factory coordination and PAC solutions typically appear as layers within a more complete MRP II architecture.
- Leitstand systems offer a more user-friendly method of visualizing and manipulating schedules at the execution planning layer of an MRP II type planning framework. Their graphical representation can be very useful for the analysis of heavily loaded resources and allow for human intervention to add more intelligence to the scheduling process. The systems expect to receive planned orders as input from an MRP-like system, and their role is to sequence these in a reasonable fashion and release them as manufacturing orders to the shop floor. In effect, this represents a kind of suboptimization that takes place after the MRP processor has exploded the requirements determined by the MPS. The approach is only suitable for companies that operate in a material constrained environment. Capacity constrained environments require finite resource scheduling tools in which the explosion process takes place at the same time as the scheduling process.
- Finally, there is no standard planning approach. Standard modules do exist, but these have to be combined into a suitable planning framework that fits the requirements of the considered company. An example could be the careful combination/integration of MRP II and finite resource scheduling logic to resolve the complex planning and control requirements for (semi)process industries.

References

Adelberger, H. and Kanet, J. (1991) The Leitstand – a new tool for computer-integrated manufacturing. *Production and Inventory Management Journal*, **32**, 43–48.

Bauer, A., Bowden, R., Browne, J., Duggan, J. and Lyons, G. (1991) *Shop Floor Control Systems – From Design to Implementation*, Chapman & Hall, London.

Bertrand, J.W.M., Wortmann, J.C. and Wijngaard, J. (1990) *Production Control – A Structured and Design Oriented Approach*, Elsevier, Amsterdam.

Blackstone, J. (1989) *Capacity Management*, South-Western Publishing, Cincinnati, Ohio.

Browne, J. (1988) Production activity control – a key aspect of production control. *International Journal of Production Research*, **26**, 415–427.

Browne, J., Harhen, J. and Shivnan, J. (1988) *Production Management Systems: A CIM Perspective*, Addison-Wesley, Wokingham.

Browne, J., Sackett, P. and Wortmann, H. (1994) The extended enterprise – a context for benchmarking, paper presented at the *IFIP Workshop on Benchmarking Theory and Practice*, Trondheim, Norway.

Burbidge, J.L. (1989) *Production Flow Analysis For Planning Group Technology*, Oxford Science Publications.

De Meyer, A., Nakane, J., Miller, J.G. and Ferdows, K. (1989) Flexibility: the next competitive battle. *Strategic Management Journal*, **10**, 135–144.

Dupas, F. and Higgins, P. (1993) *From Master Production Scheduling to Concurrent Enterprise Planning: Issues and Solutions*. Proceedings of CIM Europe Conference, 12–14 May, RAI Amsterdam, organized by CEC DG XIII, ESPRIT-CIME. In C. Kooij, P.A. MacConaill and J. Bastos (eds), *Realizing CIM's Industrial Potential*, 105 Press, pp 114–123.

Dupas, F. and Schepens, L. (1993) Établir un plan directeur de production á l'aide de la recherche operationelle et d'une interface fenêtreé (Making an MPS with operational research and a windowing interface). Conference: Optimisation industrielle et programmation lineare, technologies et applications (Industrial optimisation and linear programming, technologies and applications) organized by AFCET, 17 November. In Proceedings of the AFCET Conference, Paris, pp 49–60.

Eureka, W.E. and Ryan, N.E. (1988) *The Customer Driven Company*, ASI Press, Michigan.

Falster, P., Rolstadas, A. and Wortmann, H. (1991) FOF Production theory: toward an integrated theory for the design, production and production management of complex, one of a kind products in the factory of the future, ESPRIT Basic Research Action 3143, work package 2 report.

Goldratt, E. (1990) *The Haystack Syndrome, Sifting Infromation Out of the Data Ocean*, North River Press, New York.

Goldratt, E. and Cox, J. (1986) *The Goal*, Creative Output Books.

Hayes, R.H., Wheelwright, S.C. and Clark, K.B. (1988) *Restoring Our Competitive Advantage*, Free Press, New York.

Higgins, P. and Browne, J. (1992) Master production scheduling: a concurrent planning approach. *International Journal of Production Planning and Control*, **3**, 2–18.

IMPACS Consortium (1992) ESPRIT Project No. 2338, Integrated Manufacturing Planning and Control Systems (IMPACS) – Milestone 6 Deliverables, Deliverable to the European Commission.

Jones, G. and Roberts, M. (1990) *Optimized Production Technology*, IFS publications.

Jordan, P., Browne, J. and Browne, M. (1993) *Production Activity Control for Small Manufacturing Enterprises*. Proceedings of IFIP International Workshop on Knowledge-based Reactive Scheduling, Athens. 1 October, Elsevier Science, N.H.

Köhler, C. (1993) New technological and organizational solutions for shopfloor control. *Computer Integrated Manufacturing Systems*, **6**(1).

Martin, A.J. (1990) *Distribution Resource Planning*, Oliver Wight Publications, Essex Junction,Vermont.

Mintzberg, H. (1983) *Structures in Five; Designing Effective Organizations*, Prentice-Hall International.

Orlicky, J. (1974) *Materials Requirements Planning: the New Way of Life in Production and Inventory Management*, McGraw-Hill, New York.

Plenert, G. and Best, T.D. (1986) MRP, JIT and OPT – what's best? *Production and Inventory Management*, 2nd Quarter, 22–28.

Porter, M.E. (1980) *Competitive Strategy: Techniques for Analysing Industries and Competitors*, Free Press, New York.

Porter, M.E. (1985) *Competitive Advantage*, Free Press, New York.

Quinn, J.B., Mintzberg, H. and James, R.M. (1988) *The Strategy Process*, Prentice-Hall.

Schoenberger, R.J. (1986) *World Class Manufacturing*, Free Press, New York.

Umble, M. and Srikanth, M.L. (1990) *Synchronous Manufacturing, Principles for World Class Excellence*, South-Western Publishing, Cincinnati, Ohio.

Van Veen, E.A. (1990) Modelling products structures by generic bills-of-material, PhD thesis, Eindhoven University of Technology, The Netherlands.

Vollmann, T.E., Berry, W.L. and Whybark, D.C. (1988) *Manufacturing Planning and Control Systems*, 2nd edn, Dow Jones-Irwin, Homewood, Illinois.

Wemmerlov, U. (1984) Assemble-to-order manufacturing: implications for materials management. *Journal of Operations Management*, **4**, 347–368.

Wight, O.W. (1981) *MRP II: Unlocking America's Productivity Potential,* CBI Publishing, Boston.

Wortmann, J.C. (1990) *Towards One-Of-A-Kind Production: The Future Of European Industry.* Proceedings of the International Conference for Advances in Production Management Systems, October, Espoo, Finland.

Index